Praise for *Systems Thinking for Social Change*

"It is not hard for people to appreciate that fragmented, piecemeal efforts to solve complex problems are ineffective. But having concrete approaches to an alternative is another matter. After almost four decades of applying practical systems-thinking tools in diverse settings, David Stroh has produced an elegant and cogent guide to what works. Research with early learners is showing that children are natural systems thinkers. This book will help to resuscitate these intuitive capabilities and strengthen them in the fire of facing our toughest problems."

—**Peter Senge,** senior lecturer, MIT, founding chair, Academy
for Systemic Change, and author of *The Fifth Discipline*

"*Systems Thinking for Social Change* uses clear, down-to-earth language to explain and illustrate systems thinking, why it matters, and how it can lead to greater success in the social sector. The book is brief yet deep, big picture yet rigorously analytical. Stroh displays considerable narrative skill, especially when he shares numerous stories from his practice of applying various systems tools that led groups to new and startling conclusions. Reading this book will test the reader substantially, as the author invites us to a deeper level of introspection about our own role in systems failures of every kind—organizational and societal—and gently asks us to embrace a new way, not merely of thinking but of being in the world. A remarkable book."

—**David Nee,** Growth Philanthropy Network, and former
executive director, William Caspar Graustein Memorial Fund

"Drawing on a deep well of experience, Stroh masterfully weaves metaphor, story, and practical tools, modeling for us all effective systems thinking in action. Read it and get ready to take your game up a notch."

—**Linda Booth Sweeney,** author of *Connected Wisdom*
and coauthor of the *The Systems Thinking Playbook*

"David Peter Stroh has been a pioneer in the effort to bring principles of systems into the service of those striving for constructive social change. (I took a course from him over thirty years ago.) Many books tell you how to engage in systems thinking but not how to apply it. This is a very useful exception. Peter draws on many years of professional engagement with the important problems of our society. Of course reading his book won't let you banish all those problems. But it will help you focus your effort where you can have the best impact, and it will show you how to enlist others in the effort."

—**Dennis Meadows,** coauthor of *Limits to Growth* and
former director, Institute for Policy and Social Science

"The philanthropic sector has shifted from a 'charity' mindset to a focus on changing systems to create sustainable change. *Systems Thinking for Social Change* offers practical tools for those serious about improving communities and organizations. It doesn't minimize the complexity, but rather empowers social-change agents with tools to understand the complexity and identify the leverage points."

—**Teresa Behrens,** director, Institute for Foundation and Donor Learning

"If there is only one book you read on systems thinking, it should be *Systems Thinking for Social Change*. If you're new to systems thinking, I consider this a must read. If you've been involved in systems thinking for some time and want a renewed and extended perspective, I highly recommend it. Stroh's new work covers all the relevant areas appropriate for a solid introduction to systems thinking, though it doesn't stop there. It makes a serious contribution by detailing a number of real-world situations that have been investigated and improved using the approach presented in the book. And it does very well something that I've not seen done before: it not only shows how to map the current system, but also shows how to then create a revised map of how the system is intended to work in the future. This approach ends up identifying where measurements should be made on an ongoing basis to ascertain whether the system is undergoing the intended transformation."

—**Gene Bellinger,** director, Systems Thinking World, Inc.

"With this book Stroh has produced an essential—and long overdue—guide to applied systems thinking. A few well-selected examples of initiatives that turned from 'working' to 'transformative' lay the foundation for how change makers can address chronic, complex social problems and deepen their impact. After helping the reader recognize what might be holding their interventions back, the book moves with ease into ways of finding leverage, the use of systems stories, and the power of visioning. In Stroh's capable hands, systems thinking becomes a tool for defining personal or organizational priorities, for planning, and for evaluating success through measurable indicators. But the book is much more than a formidable toolbox from which to draw on a daily basis. It is, at its deeper level, a warm invitation to cultivate systems thinking as 'a way of being, not just doing, so that on the way to long-lasting, desirable outcomes, change makers can become more and more the change they want to see."

—**Marta Ceroni,** executive director, Donella Meadows Institute

"David Stroh, in his invaluable new book, shows that good intentions are not enough for those who aspire to make lasting progress on fundamental social issues—and also how the language and tools of systems theory can provide a deeper understanding of the root causes and help identify the leverage points for productive and sustainable change."

—**Russell Eisenstat,** executive director,
Center for Higher Ambition Leadership

"Over fifteen years ago, David Stroh was instrumental in introducing systems thinking to the peace-building field, using tools that have proven to be powerful for improving the effectiveness of our work. This book is a valuable resource for our field—a must read for all practitioners who have been seeking practical and easy-to-understand guidance on using systems thinking for conflict analysis and strategic planning."

—**Diana Chigas,** professor of practice, Fletcher School
of Law and Diplomacy at Tufts University, and co-director of
collaborative learning, CDA Collaborative Learning Projects

"Stroh has offered an important gem in his new book, *Systems Thinking for Social Change.* Both illuminating and immediately useful, the book shares the key dynamics and success factors gleaned from his long career of working with organizations struggling with society's most persistent issues. A must read for anyone whose aim is to make a difference on the ground."

—**Kristina Wile,** co-president, Leverage Networks,
and managing partner, Systems Thinking Collaborative

"This is a must read for public leaders and citizens who are interested in the learning disciplines required for a sustainable, proactive approach to preserving our shared resources."

—**Georgianna Bishop,** president, The Public Sector Consortium

"For those who have worked for many years in the social-service sector and who have grown cynical or disillusioned as to whether it is even possible to effect major social change, David Peter Stroh's book, *Systems Thinking for Social Change,* is a must read—a clear, thoughtful, and practical guide for those desiring to create lasting social change. But reader beware! Systems thinking is more than a new way of thinking. As Stroh puts it, it is a new way of being. It requires the ability to look at things in a new way, to interact with others differently, to have a clear vision of where you want to go, a willingness to see things the way they are, and, finally, the courage to take responsibility for why the system as is isn't working. If you want to help create long-lasting, effective social change, if you want to say 'we're doing it—we're actually making progress,' then read this book."

—**Anne Miskey,** executive director, Funders Together to End Homelessness

"As philanthropic organizations increasingly seek to strengthen their impact, the perspectives, methods, and tools described in Stroh's book provide us with critical guidance for thinking and action to address complex social problems and for building 'all-in' approaches to problem solving. Anyone in government, nonprofits, or philanthropy can benefit from this approach to solutions. And while it might take a lifetime to master the use of systems thinking for social change, reorienting how we think about problems in this way can immediately set us on a new path toward sustainability and greater likelihood of success."

—**Lexi Nolen,** vice president, Episcopal Health Foundation

"Systems thinking quickly gets very abstract and technical, often underplaying the social and storytelling dimensions. For a long time I've been looking for a more practical, readable, and engaging introductory book for my classes. Now, finally, here it is!"

—**Per Espen Stoknes,** author of *What We Think About When We Try Not To Think About Global Warming* and senior lecturer at BI Norwegian Business School

"Societal problems are a swirl of causes, effects, interactions, and contributing relationships. Yet, too often, simplistic answers are applied by the well-intended that only touch on one strand of what is (in reality) a complex and interconnected web. Stroh's work provides an actionable guide on how to model these relationships—and more importantly how to have a meaningful and lasting impact on them."

—**Jason E. Glass,** superintendent and chief learner, Eagle County Schools

SYSTEMS THINKING

for

SOCIAL CHANGE

A Practical Guide to Solving Complex Problems,
Avoiding Unintended Consequences,
and Achieving Lasting Results

DAVID PETER STROH

Chelsea Green Publishing
White River Junction, Vermont

Editor: Joni Praded
Project Manager: Bill Bokermann
Copy Editor: Laura Jorstad
Proofreader: Helen Walden
Indexer: Peggy Holloway
Designer: Melissa Jacobson

Printed in the United States of America.
First printing September, 2015.
10 9 8 7 6 5   18 19 20 21

Our Commitment to Green Publishing
Chelsea Green sees publishing as a tool for cultural change and ecological stewardship. We strive to
align our book manufacturing practices with our editorial mission and to reduce the impact of our
business enterprise in the environment. We print our books and catalogs on chlorine-free recycled
paper, using vegetable-based inks whenever possible. This book may cost slightly more because it
was printed on paper that contains recycled fiber, and we hope you'll agree that it's worth it. Chelsea
Green is a member of the Green Press Initiative (www.greenpressinitiative.org), a nonprofit coalition of publishers, manufacturers, and authors working to protect the world's endangered forests
and conserve natural resources. *Systems Thinking for Social Change* was printed on paper supplied
by McNaughton & Gunn that contains 100% postconsumer recycled fiber.

Library of Congress Cataloging-in-Publication Data
Stroh, David Peter, 1950-
Systems thinking for social change : a practical guide to solving
complex problems, avoiding unintended consequences, and achieving
lasting results / David Peter Stroh.
    pages cm
ISBN 978-1-60358-580-4 (paperback) -- ISBN 978-1-60358-581-1 (ebook)
1. Social change. 2. Management. 3. System theory. I. Title.
HM831.S787 2015
303.4--dc23
                              2015022577

Chelsea Green Publishing
85 North Main Street, Suite 120
White River Junction, VT 05001
(802) 295-6300
www.chelseagreen.com

FSC
www.fsc.org
MIX
Paper from
responsible sources
FSC® C011935

Dedicated to
My grandfather, Dr. Albert Sondheimer,
and mother, Eva Sondheimer Stroh,
Who embodied the commitment to repairing the world

CONTENTS

PART THREE
Shaping the Future

INTRODUCTION

Whether you are committed to ending homelessness, strengthening education, improving public health, reducing the problems of poverty, developing environmental sustainability, or helping people live better lives in other ways, you might have noticed that the organizations and systems you want to change have a life of their own. In other words, you do things to try to improve them and they essentially continue to operate as if your input makes no difference.

Organizations and social systems do in fact have a life of their own.

As someone committed to achieving sustainable, breakthrough social change, it helps to understand these forces so that you can consciously work *with* them instead of unconsciously working against them. You might be working in a foundation or nonprofit, government agency or legislature, department of corporate social responsibility, or as a consultant to people in these roles. In an era when growing income inequality and climate change increase the vulnerability of many and reduce the sustainability of all, you might feel called to do more to heal the world. You might also be challenged to achieve more with less—less time, attention, and money than you had before.

The book is based on a simple premise: *Applying systems thinking principles and tools enables you to achieve better results with fewer resources in more lasting ways.* Systems thinking works because it:

- Increases your awareness of how you might unwittingly be contributing to the very problems you want to solve.
- Empowers you to begin from where you can have the greatest impact on others, by reflecting on and shifting your *own* intentions, thinking, and actions.
- Mobilizes diverse stakeholders to take actions that increase the effectiveness of the whole system over time instead of meeting their immediate self-interests.

- Helps you and others anticipate and avoid the negative longer-term consequences of well-intentioned solutions.
- Identifies high-leverage interventions that focus limited resources for maximum, lasting, systemwide improvement.
- Motivates and supports continuous learning.

More specifically, if you are the director or program officer in a foundation, learning to think systemically will help you be more effective in your roles as convener, grant maker, and educator/advocate. You will become a better convener by:

- Enabling diverse stakeholders to see the big picture.
- Catalyzing conversations of accountability among stakeholders so that each becomes aware of how they unwittingly contribute to the very problem they want to solve.
- Mobilizing people to optimize the whole system instead of just their part of it.

You can increase your effectiveness as a funder by using systems thinking to:

- Uncover root causes of chronic, complex problems.
- Identify high-leverage interventions.
- Strengthen your commitment to invest for the long term and your ability to evaluate impact over time.

You can become a more effective educator and advocate by using systems thinking to:

- Inform policy makers and the public about the short- versus long-term consequences of proposed solutions.
- Reduce people's addictions to quick fixes that are likely to only make matters worse in the long run.
- Champion early small successes that also support people's higher and longer-term aspirations.

If you are a nonprofit or government agency that depends on outside funding, you can use systems thinking to:

- Deepen your understanding of the problems you want to solve.
- Engage people in the communities you serve more effectively.

- Distill your insights into visual systems maps that "are worth a thousand words."
- Identify strategic interventions that best leverage limited resources.
- Write more powerful grant requests that incorporate all of the above.

As a legislator or policy maker, you can use systems thinking to:

- Think more clearly about why social problems persist despite your best efforts to solve them.
- Anticipate and avoid long-term negative unintended consequences of proposed solutions.
- Identify high-leverage interventions that make the most of public tax dollars.
- Powerfully communicate the benefits of proposed legislation and policy to your constituents based on all of the above.

You can use systems thinking as a corporate social responsibility manager to develop more effective partnerships with key external stakeholder groups in both the nonprofit and public sectors. It can help you:

- See the big picture more clearly.
- Uncover and own the unintended negative consequences of your own actions.
- Work with external stakeholders to develop solutions that are more likely to benefit all parties over time.

If you are a professional in the systems thinking community who is committed to social change, you can learn how to integrate the tools of systems thinking into a proven change management process.

If you are an organizational or community development consultant, you can use systems thinking to increase people's motivations to change, facilitate collaboration across diverse stakeholders, identify high-leverage interventions, and inspire a commitment to continuous learning.

What You Will Learn

Systems Thinking for Social Change helps you achieve these benefits by understanding what systems thinking is and how it can empower your work. It will also help you appreciate the basic principles and tools of systems thinking, and learn how to apply it to problem solving, decision making, and strategic planning *without* becoming a technical expert.

More specifically, you will learn to:

Use systems thinking instead of more conventional linear thinking to address chronic, complex social problems. Einstein observed, "The significant problems we face cannot be solved with the same level of thinking we were at when we created them." Systems thinking is more appropriate than conventional thinking to solve chronic, complex social problems. By contrast, you can unwittingly perpetuate such problems by thinking conventionally about how to solve them.

Apply systems thinking as both a set of principles and a particular group of analytic tools. The tools in *Systems Thinking for Social Change*—which include the iceberg, the causal loop diagramming and systems archetypes popularized by Peter Senge, and the Bathtub Analogy—have proven highly effective in shifting how people address social problems.[1] While many other analytic tools exist,[2] this book demonstrates why these specific tools are especially helpful in enabling a diverse group of stakeholders to, in the words of executive consultant Ram Charan, "cut through complexity to the heart of the matter, without being superficial."[3]

Integrate systems thinking into a proven four-stage change management process. There are many change processes that seek to align diverse stakeholders without helping people understand how their thinking and subsequent behavior unintentionally undermines their own per-formance, the performance of others in the system, and the system's effectiveness as a whole. In other words, they often establish common ground around a shared aspiration yet fail to help people develop a joint understanding of not only what has been happening but also *why*. In searching for root causes, people typically assume that they are doing the best they can and that someone else is to blame—instead of recognizing, in the words of leadership expert Bill Torbert, that "if you are not aware of how you are part of the problem, you can't be part of the solution." By contrast, systems thinking enables people to identify high-leverage interventions based on deep insights into root causes that incorporate their own thinking and behavior.

This book reveals a four-stage change management process, grounded in systems thinking, that my longtime colleague Michael Goodman of Innovation Associates Organizational Learning and I have been working with for more than fifteen years. It also discusses how you can build systems thinking into other change processes. Many new processes have emerged in recent years to engage diverse stakeholders as a way of managing complexity and sharing resources.[4] From a systems thinking perspective, *the key is to help participants cultivate a deep awareness of current reality as something they have created instead of as something that exists outside of and independent of them.*

Catalyze an explicit choice between the purpose people say they want to accomplish and the benefits they are achieving right now. Systems are perfectly designed to achieve the results they are currently achieving.[5] In other words, no matter how dysfunctional a system appears to be, it is producing benefits for the people who participate in it. A pivotal intervention you will learn in this change process is to help people compare the benefits of change with the benefits of the status quo—and then help them make a conscious choice between the payoffs they are now getting and the espoused purpose they say they want the system to accomplish. This involves deepening people's connections with what they care about most and supporting them to let go of current payoffs that do not serve their highest aspirations.

Apply systems thinking prospectively as well as retrospectively. The book highlights the application of systems thinking *retrospectively* to help people develop better solutions to chronic, complex social problems by first deepening their understanding of why they have been unsuccessful so far despite their best efforts. Emphasizing the retrospective application of systems thinking is so important because people tend to create more problems by failing to first fully appreciate the problem they are trying to solve.

At the same time, the book also shows you how to use the tools *prospectively* for strategic planning and assessment. You will learn to integrate leverage points into a systemic theory of change, design new systems where there is no precedent, organize your priorities, and establish an evaluation method grounded in systems principles.

Cultivate systems thinking as a way of being—not just as a way of thinking. Because systems thinking challenges people to take more responsibility for their actions and make hard choices, it is framed in this book as *more* than a way of *thinking*. The book describes how the approach affects people not only cognitively but also emotionally, spiritually, and behaviorally. As you build your capacity to think systemically, you will discover that the tools both enable and require you to develop a new way of *being*, not just *doing*—a set of character traits to cultivate (such as curiosity, compassion, and courage) that complement and deepen your new skills.

The concepts will be tied closely to experiences my colleagues and I have had in applying systems thinking to social change initiatives. Some of the stories you will read about address:

- Aligning a community of a hundred thousand people around a ten-year plan to end homelessness.
- Designing a more effective statewide early-childhood development and education system.
- Improving the quality of environmental public health in states, counties, and cities around the United States.
- Reforming the criminal justice system with particular attention to reducing recidivism among people recently released from prison.
- Improving relationships between two agencies responsible for improving K–12 education in their state.
- Increasing people's fitness and consumption of healthy local food in a rural region.

Structure of the Book

The book is organized into three parts. Part 1 introduces systems thinking within the context of social change and includes four chapters. Chapter 1 explains why people's best intentions to solve chronic social problems often fail to deliver expected results, defines *system* and *systems thinking*, and distinguishes systems thinking from conventional thinking. Chapter 2 explains why systems thinking is effective in meeting four challenges of managing change, identifies six indicators that help you determine when applying systems thinking is likely to be most useful, and describes in particular how it can contribute to the pioneering cross-sector coordination process known as Collective Impact.

Chapter 3 introduces the metaphor of systems thinking as storytelling. It distinguishes two types of stories: a more common one that tends to perpetuate the status quo, and a systemic one that stimulates productive change. This chapter also explains the power of language to create stories and summarizes the basic elements for formulating a systems story. Chapter 4 deepens this metaphor by illuminating basic story lines and richer systemic patterns or plots that underlie a diverse set of social issues. If you are already familiar with systems thinking, you might want to pay particular attention to the ways in which systems thinking mobilizes change (chapter 2) and skim the next two chapters for their many social and environmental examples.

Part 2 describes the four-stage change process. Chapter 5 introduces this four-stage process as:

1. Building a foundation for change.
2. Seeing current reality more clearly.
3. Making an explicit choice about what is most important.
4. Bridging the gap between people's aspirations and current state.

Chapter 6 describes how to build a foundation for change by identifying and engaging key stakeholders, establishing common ground, and developing collaborative capacity. It addresses such challenges as working with stakeholders who are motivated by immediate self-interests as well as higher aspirations, focusing their efforts around what can seem like a boundless challenge, and building relational skills that enable people to become responsible participants in a complex world.

Chapter 7 explains how stakeholders can take a deep dive into current reality early in the change process—and also why this step is critical. Often, people begin a social change process with similar aspirations for the outcome but have very different perceptions of what the real difficulty is and what should be done to solve it. They don't appreciate how their own intentions, beliefs, and behavior affect the performance of others as well as their own. Failing to see the big picture, they are more likely to propose familiar solutions that risk perpetuating the very problem they have been trying to solve. In other cases, people might feel overwhelmed by the complexity of the big picture and question if they can do anything differently. The chapter helps readers assess current reality by recommending how to gather and organize information for a systems analysis; presenting the results of systems analyses from several social change initiatives; and showing how to create systems analyses that are comprehensive enough to cover many critical factors and viewpoints, yet simple enough to communicate and act upon.

Chapter 8 addresses how to build support for the outcome of a systems analysis. Helping stakeholders accept new insights about the system warrants a separate chapter because the language of systems thinking is often unfamiliar and the message of shifting from blame to responsibility can be difficult to embrace. The chapter describes three ways to meet these challenges: engaging people to develop their own analysis as much as possible; surfacing the mental models that influence how people behave; and creating catalytic conversations that stimulate awareness, acceptance, and alternatives.

Chapter 9 guides stakeholders to make an informed and explicit choice about the purpose they *want* the system to accomplish. Since a system is always designed to achieve *something*, no matter how dysfunctional it seems, a pivotal intervention is to help people distinguish, and where necessary choose, between what the system is accomplishing right now and what they aspire to accomplish. The chapter provides a way to help people compare the case for change (in line with what they deeply want) with the case for the status quo (the often hidden benefits people get from participating in the system as it behaves now). Because there are usually trade-offs between current benefits and people's higher aspirations, the process supports stakeholders to make an explicit choice based on what they care about most deeply, feel most drawn to achieve, and are willing to let go of to achieve what is even more important to them. At this point shared visioning takes on more meaning, since people also acknowledge the sacrifices that realizing the vision entails. They become more willing to optimize the whole system instead of continuing to optimize just their part.

Chapter 10 helps people bridge the gap between current reality and their consciously chosen direction. It involves identifying leverage points and establishing a process of continuous learning and outreach. The chapter describes four generic leverage points: increasing people's awareness of their often non-obvious interdependencies; "rewiring" key cause–effect relationships; shifting underlying beliefs and assumptions; and aligning the chosen purpose with updated goals, metrics, incentives, authority structures, and funding streams. It also shows how to extend the process begun in the previous four stages through learning from experience, expanding the resource pool, and scaling up what works.

Part 3 describes three ways you can use systems thinking to shape the future. Chapter 11 addresses how to apply systems thinking to strategic planning. It explains the advantages of systemic over linear theories of change; introduces two core systemic theories of change; and shows how these theories can be used integrate leverage points, critical success factors, and disparate priorities to develop coherent and navigable strategies over time. Chapter 12 answers the question frequently posed by private and public funders about how systems thinking can contribute to evaluation. The chapter provides broad guidelines for conducting systemic evaluations and specific recommendations for validating explicit systemic theories of change.

Finally, chapter 13 provides guidelines for developing your ability as a systems thinker over time. It offers three ways forward: cultivating an orientation that integrates the cognitive, emotional, behavioral, and spiritual dimensions of the systems approach; learning by doing; and asking systemic questions.

In summary, the book provides you with many ways to think and act more productively, using methods that have been tested over decades with clients facing a wide range of seemingly intractable social problems. The tactics in the pages ahead can help even highly skilled change makers get closer to their goals and develop some crucial lifelong problem-solving traits.

For several years, I taught in a national program that introduced systems thinking approaches to fellows in the Centers for Disease Control's Environmental Public Health Leadership Institute. A report conducted on the program indicated that fellows learned to:

- Think through difficult issues.
- Understand what they do not know and how to learn it.
- Ask great questions.
- Engage others more effectively by seeing reality from their perspective.
- Apply a problem-solving approach that is both flexible and concrete.

- See a bigger picture that clarifies connections among many factors and identifies root causes of complex problems.
- Focus on what is most important.
- Work toward deep systems change by transforming the underlying assumptions and policies that govern existing processes and procedures.

These results represent the promise of systems thinking for social change and the opportunities available to you as you read on.

PART ONE

SYSTEMS THINKING FOR SOCIAL CHANGE

CHAPTER 1

Why Good Intentions
Are Not Enough

Consider the following headlines, which are all based on true stories:

HOMELESS SHELTERS PERPETUATE HOMELESSNESS

DRUG BUSTS INCREASE DRUG-RELATED CRIME

FOOD AID INCREASES STARVATION

"GET-TOUGH" PRISON SENTENCES FAIL TO REDUCE
THE FEAR OF VIOLENT CRIME

JOB TRAINING PROGRAMS INCREASE UNEMPLOYMENT

What is going on here? Why do seemingly well-intentioned policies produce the opposite of what they are supposed to accomplish?

If you look closely at these solutions and many other stories of failed social policies, they have similar characteristics. They:

- Address symptoms rather than underlying problems.
- Seem obvious and often succeed in the short run.
- Achieve short-term gains that are undermined by longer-term impacts.
- Produce negative consequences that are unintentional.
- Lead us to assume that we are not responsible for the problem's recurrence.

For example, get-tough prison sentences do not address the socioeconomic causes of most inner-city crime. Although the perpetrators go to prison and pose less of an immediate threat, 95 percent of them are eventually released back into society—hardened by their experience and ill prepared to reenter their communities productively. Nearly half of those released from prison are imprisoned again within the first three years for committing a repeat offense.[1] Moreover, the current system further weakens the infrastructure of these communities because it incarcerates fathers and mothers who can no longer bring up their children—thereby creating more instability and increasing the likelihood of producing a new generation of people who commit crimes. The system also redirects valuable public funds away from the socioeconomic and criminal justice reforms that could reduce crime permanently. Finally, if a formerly incarcerated person commits another crime, he or she is sent back to prison without considering how get-tough policies might have contributed to the recidivism.

Lewis Thomas, the award-winning medical essayist, observed, "When you are confronted by any complex social system . . . with things about it that you're dissatisfied with and anxious to fix, you cannot just step in and set about fixing with much hope of helping. This is one of the sore discouragements of our time."[2] He went on to say, "If you want to fix something you are first obliged to understand . . . the whole system."

Distinguishing Conventional from Systems Thinking

What does it mean to understand the whole system? First, it means appreciating a situation you want to change through a systemic instead of a conventional lens. If you think that a systems lens is too sophisticated and beyond most people's reach, let me assure you that it is child's play.

If you are a parent, remember when your children were young, and you picked up after them. Your children would let their clothes pile up on the floor and move on to something more interesting. Eventually, after numerous failed attempts to have them put their clothes in the laundry, you would give up and put them there yourself. When your children would come back, the clothes had disappeared—as if by magic. "That worked!" they concluded. Nonlinear cause and effect, time delay, success (from their point of view, not necessarily yours)—these are all signs of a highly competent systems thinker.

Conventional or linear thinking works for simple problems, such as when I cut my hand and put on a Band-Aid to help the cut heal. It is also the basis for how most of us were taught in school and still tend to think—divide the world into specific disciplines and problems into their components under the assumption that we can best address the whole by focusing on and optimizing the parts.

However, conventional thinking is not suited to address the complex, chronic social and environmental problems you want to solve. These problems require systems thinking, which differs from conventional thinking in several important ways, as table 1.1 shows.

For example, if the problem is homelessness, then the solution is not simply providing shelter. Providing temporary shelter is insufficient since people tend to cycle through shelters, the street, emergency rooms, and jails. Moreover, it is too easy to conclude that when people remain homeless, they do not want their own place to live, when in fact many want the security that comes with permanent housing. In addition, funding shelters tends to undermine both the political will and financial resources required to end homelessness.

Ending homelessness requires a complex, long-term response involving affordable permanent housing, support services for the chronically homeless, and economic development. This means establishing new relationships among the various providers who prevent homelessness, those who help people cope with being homeless, and those who develop the permanent housing with support services and jobs that enable people to

TABLE 1.1. CONVENTIONAL VERSUS SYSTEMS THINKING

Conventional Thinking	Systems Thinking
The connection between problems and their causes is obvious and easy to trace.	The relationship between problems and their causes is indirect and not obvious.
Others, either within or outside our organization, are to blame for our problems and must be the ones to change.	We unwittingly create our own problems and have significant control or influence in solving them through changing our behavior.
A policy designed to achieve short-term success will also assure long-term success.	Most quick fixes have unintended consequences: They make no difference or make matters worse in the long run.
In order to optimize the whole, we must optimize the parts.	In order to optimize the whole, we must improve *relationships* among the parts.
Aggressively tackle many independent initiatives simultaneously.	Only a few key coordinated changes sustained over time will produce large systems change.

Source: Innovation Associates Organizational Learning

end homelessness. Aligning providers along this continuum of care toward a goal of affordable permanent housing with support services increases everyone's ability to solve the problem.

The principle that solutions that work in the short run often have negative long-term effects, a phenomenon known as better-before-worse behavior, has significant implications for funders and policy makers. It raises what ·foundations call the philanthropic challenge—the task of determining how to fix a problem now versus help people over time. It also challenges public policy makers and business leaders to educate their constituents (such as private citizens and financial investors) about the risks of alleviating short-term pressures and fears without understanding the potential negative consequences of expedient solutions. In a world that promotes instant gratification, it can be difficult to remind people that "there is no such thing as a free lunch."

This contrasting principle is known in systems terms as worse-before-better behavior. This means that long-term success often requires short-term investment or sacrifice. If you want to motivate people to work toward long-term success, then you as a leader must act in accordance with your own highest, long-term aspirations. The principle challenges leaders to:

- Resist quick fixes that actually undermine long-term effectiveness.
- Set realistic expectations with the people they serve.
- Target short-term successes that deliberately support long-term results and provide people with true hope instead of false promises.

Refining the Definition of Systems Thinking

Another useful distinction to introduce here is the difference between a system and systems thinking. The award-winning systems thinker Donella Meadows defined a system as "an interconnected set of elements that is coherently organized in a way that achieves *something* [italics mine]."[3] Meadows's definition points to the fact that systems achieve a purpose— which is why they are stable and so difficult to change. However, this purpose is often not the one we *want* the system to achieve.

Building on her definition, I define systems thinking as the ability to understand these interconnections in such a way as to achieve a *desired* purpose. One of the benefits of systems thinking is that it helps people understand the purpose that a system is accomplishing. This prompts them to reflect on the difference between what they say they want (their espoused purpose) and what they are actually producing (their current purpose). Reconciling the difference between these two is the subject of chapter 9.

As a reader you may have come across different schools of systems thinking, such as general systems theory,[4] complexity theory, system dynamics, human system dynamics, and living systems theory. It is helpful to recognize that, while all these schools tend to agree on most of the systems principles described in table 1.1, they differ in the methodologies used to both analyze a system and identify ways to improve it.

This book is primarily based on the concept of causal feedback loops in systems, and uses the causal feedback loop diagramming tools first popularized in Peter Senge's management classics *The Fifth Discipline* and *The Fifth Discipline Fieldbook*.[5] These tools can be integrated with other kinds of systems analysis, such as system dynamics and soft systems methodology.

I emphasize the feedback tools for several reasons. As a co-founder with Charlie Kiefer, Peter Senge, and Robert Fritz of Innovation Associates, the consulting firm that pioneered many of the ideas referred to in Peter's book, I have thirty-five years of experience in working with these tools. I have also seen how powerful they can be in achieving sustainable, breakthrough change in the nonprofit, public, and private sectors—as well as in communities that engage leaders from all three sectors. Moreover, the tools are broadly recognized (as evidenced by the popularity of the work that Peter and many others among our colleagues have done) and readily understood by a wide range of people.

By introducing systems thinking as "the Fifth Discipline" of creating a learning organization, Peter Senge embedded what might otherwise be a purely technical and cognitive set of tools into a broader context. This context embraces multiple dimensions:

- **Spiritual:** The ability to see and articulate what will benefit diverse people over time.
- **Emotional:** The ability to master our emotions in service of a higher purpose.
- **Physical:** The ability to bring people together and enable them to collaborate.
- **Mental:** The ability to recognize how our individual and collective thinking affects the results we want.

This last point illustrates another critical benefit of this methodology, which is its emphasis on responsibility and empowerment. Every day, we can look around and see unintended consequences arising from what seemed at one time to be best-laid plans. Undoubtedly, whoever cast those plans had the best of intentions. A judge incarcerating parent after parent might have thought he was protecting citizens, but may not have fully understood he was also exacerbating problems for the children left behind, and perpetuating criminal behavior over time. The director of a shelter might have thought that she was protecting homeless people from the elements, but may not have fully understood that shelters divert critical resources from the even more humane and sustainable solution of permanent supportive housing.

In other words, burdens are shifted, unexpected results surface, and a host of other systems issues arise from good intentions. For any complex problem to be solved, the individual players all need to recognize how they unwittingly contribute to it. Once they understand their own responsibility for a problem, they can begin by changing the part of the system over which they have the greatest control: themselves. As we'll see in the pages ahead, the greatest opportunities for lasting change arise when all the players reflect on and shift their own intentions, assumptions, and behavior.

Closing the Loop

- People's good intentions to improve social systems are often undermined when they apply conventional thinking to chronic, complex social problems.
- Systems thinking is different from conventional thinking in several important ways.
- A social system is always designed to achieve a purpose, although it might not be the purpose people say they want it to accomplish.
- The specific approach to analyzing a complex social system that is used in this book has many important benefits, and it can also be used to complement other analytic methods.

CHAPTER 2

Systems Thinking Inside: A Catalyst for Social Change

In the summer of 2011 a group of leaders representing the Iowa Department of Education (IDE) and the state's regional Area Education Agencies (AEAs) met to improve how they worked together. The two organizations had historically been funded separately and operated independently even though both were responsible for the quality of K–12 education in the state. But new challenges had arisen: Budget constraints had amplified, and test scores had failed to keep pace with increases in the national average. Iowa's kids needed a better, more integrated support structure, and this demanded that the two organizations partner more closely. Other states face similar challenges: rising educational expectations, tighter budgets, and tensions between a central organization that promotes standardized systems and districts that want to pursue innovations tailored to their immediate constituents.

Forging a partnership is not easy, as demonstrated by the many examples of failed mergers in all sectors. However, in the case of Iowa education, the new partnership Collaborating for Iowa's Kids has been highly successful. Both organizations are now operating with a shared purpose, vision, and set of values; supporting a jointly developed theory of success; working toward common goals; meeting monthly along with local school districts; reviewing shared metrics; and achieving meaningful results. Since the program began four years ago, it has expanded to include 80–100 participants. An oversight group keeps strategies and initiatives connected and assures that new initiatives are truly collaborative. Work groups including IDE, AEA, and local school district members develop priority initiatives. The larger group operates as a learning community: reviewing data, addressing unintended consequences of implementation actions, suggesting ways that implementation can be strengthened, and reviewing plans of work groups. An early literacy initiative has pilot districts throughout the state using comprehensive approaches that include professional development, protocols that serve as guides for students, school teams planning and monitoring implementation with AEA partners, assessment plans, coaches for K–3 teachers, and data collection and analysis.

What enabled this partnership to develop when so many others fall short of expectations? While there are many reasons, at least two relate to how the two groups were initially brought together. Early on, the leaders of both organizations began working with Kathleen Zurcher, an experienced consultant who continues to support the project, to define their shared aspiration—which is one anchor of a systems approach. Then the leaders applied systems thinking with me to deepen their understanding of why it might be difficult to collaborate despite their shared intentions to improve the lives of Iowa's kids.

Enabling them to develop this insight was facilitated by the fact that many chronic, complex problems can be viewed through the lens of systems archetypes—patterns of behavior that are so common within organizations, they have predictable consequences and well-understood solutions. The archetype playing out for these Iowa groups was clear: Accidental Adversaries. While each group had been conceived as part of an overall system whose actions would benefit all, each group had come to focus on its individual responsibilities and success. In the process each made life harder for the other, limiting the success of both groups and the system as a whole. For instance, the IDE had introduced many new programs to achieve its goal of providing guidance and governance to the state education system. However, these programs had disrupted the AEAs' abilities to manage their own resources, which led them in turn to either customize or disengage from the IDE initiatives, thereby making IDE's work more difficult.

The story of Accidental Adversaries resonated with both parties. It motivated them to have new conversations about how they could work as a unified system to maximize the benefits of their partnership and avoid the unintentional problems they had created for each other. The parties agreed that the role of the IDE was to set direction and lead, and the role of the AEAs was to implement. As a result, IDE moved from blaming AEAs for operating independently to working with them to meet regional needs within the statewide context, and AEAs worked to adjust their individual initiatives to fit within the statewide direction and plans. They also agreed that the local school districts (formally known as Local Education Agencies or LEAs) were integral and needed to be included in the systemwide alignment work. And finally, they committed to focusing their efforts and resources on selected priorities—beginning with the early literacy initiative.

Mark Draper, the director of special education for the Green Hills AEA, said of the initial systems analysis meeting. "This has been the most concrete and useful conversation I've had on the relationship between our two groups in the past 20 years." Connie Maxson, who was the bureau chief of teaching and learning services for the Department of Education, said, "This has been the best conversation I've had on the relationship between our two groups in the seven years since I've been here." Their understanding of the Accidental Adversaries dynamic helped build not only stronger relationships between the IDE and AEA system, but also over time between the AEA system and individual AEAs, the local school districts and the IDE, and the local school districts and individual AEAs.

How Systems Thinking Meets Four Challenges of Change

How does systems thinking help people achieve sustainable, breakthrough change? The Iowa education story—and many others—points to ways in which thinking systemically meets four common challenges of change.

First, systems thinking *motivates* people to change because they discover their role in exacerbating the problems they want to solve. For example, the IDE came to see that rolling out new programs to the AEAs without sufficiently taking their needs into account led the AEAs to customize or disengage from these programs, thereby creating inconsistent, low-quality solutions that made the IDE's own work more difficult. On the other side, the AEAs recognized that customizing or disengaging from statewide programs led the IDE to initiate even more programs that stretched the AEA's own resources.

Second, systems thinking *catalyzes collaboration* because people learn how they collectively create the unsatisfying results they experience. In the case of Iowa, both parties came to see that their localized solutions under-mined their own organization's effectiveness and children's abilities to learn. They recognized that they were in the same boat, one of their own making, and that only by working together could they design a more seaworthy craft. Based on these insights, they developed new principles and structures for partnering over time and applied systems thinking again to organize their efforts around a common theory of success.

Third, systems thinking *focuses* people to work on a few key coordinated changes over time to achieve systemwide impacts that are significant and sustainable. This approach contrasts with people's tendencies to try to do too much with too few resources and achieve less as a result. In Iowa, the organizations chose to target specific high-leverage educational outcomes beginning with early-childhood literacy because of its pivotal role in long-term student performance.

Fourth, systems thinking *stimulates continuous learning*, which is an essential characteristic of any meaningful change in complex systems. The inherent and ever-changing complexity of social problems forces people to accept that knowledge is never complete or static. Learning is a more powerful mind-set than knowing because it enables us to keep adapting in the face of new information and conditions.[1] In Iowa, the organizations put in place a process for assessing their progress and adjusting their joint strategies over time.

Table 2.1 summarizes these four change challenges and illustrates how systems thinking differs from the more common approaches to bringing diverse stakeholders together in service of social change. While many typical approaches help people recognize their shared aspirations, they often fall short, failing to show people how they are responsible for current reality.

As Peter Senge observed in *The Fifth Discipline*, "The building of shared vision lacks a critical underpinning if practiced without systems thinking." He goes on to say, "The problem lies not in shared visions themselves, so long as they are developed carefully. The problem lies in our reactive orientation toward current reality. Vision becomes a living force only when people truly believe they can shape their future. The simple fact is that most managers do not *experience* that they are contributing to creating their current reality. So they don't see how they can contribute to changing that reality."[2] When people fail to see their responsibility for the present, they (1) tend to assume that their primary work is to change others or the system—not themselves, and (2) promote solutions that optimize their part of the system based on

TABLE 2.1. MEETING THE CHALLENGES OF CHANGE THROUGH SYSTEMS THINKING

The Challenge	Benefits of a Systems Thinking Approach	Characteristics of a Conventional Approach
Motivation: Why should we change?	Show responsibility for current reality	Appeal to desire or fear
Collaboration: Why should we work together?	Demonstrate how people's current ways of interacting undermine both their individual and their collective performance	Tell people they should
Focus: What should we do?	Use leverage to change the few things that change everything else	Tackle many issues independently and simultaneously; attack symptoms
Learning: Why bother?	Recognize that our actions matter, and that we need to learn from the consequences of our actions	Assume that others are at fault and must learn

a mistaken belief that the way to optimize the whole system is to optimize each of the parts. By contrast, a systems view encourages them to critically assess their own contributions first.

When to Use Systems Thinking

Since 1991 Intel has used the motto "Intel Inside" to emphasize that its chips power computers around the world. Like those chips, systems thinking powers change within many other change management frameworks. Because it can be embedded in many different methodologies, including the one described in this book, I think of it as "Systems Thinking Inside."

It is especially effective to incorporate systems thinking into a broader systems approach when:

- ☐ A problem is chronic and has defied people's best intentions to solve it.
- ☐ Diverse stakeholders find it difficult to align their efforts despite shared intentions.
- ☐ They try to optimize their part of the system without understanding their impact on the whole.

☐ Stakeholders' short-term efforts might actually undermine their intentions to solve the problem.

☐ People are working on a large number of disparate initiatives at the same time.

☐ Promoting particular solutions (such as best practices) comes at the expense of engaging in continuous learning.

Systems thinking can also play a role in defusing unintentional conflict or opposition among stakeholders, as it did in the Collaborating for Iowa's Kids case. When conflicts run deeper and there is little willingness on the part of people to collaborate with each other based on either a common concern or shared aspiration, such as in the identity-based Israeli–Palestinian conflict, systems thinking can still help third parties better understand the underlying dynamics and identify possible interventions.[3]

The broader systems approach used in this book will be summarized in chapter 5 and detailed in chapters 6 through 10, but we will additionally consider other systems approaches people employ, such as the Collective Impact model, which can also benefit from building "Systems Thinking Inside."

Systems Thinking for Collective Impact

One of the most acclaimed approaches for large-scale social change to appear in recent years has been Collective Impact, a broad cross-sector coordination process introduced by John Kania and Mark Kramer in their pioneering *Stanford Social Innovation Review* article by the same name.[4] The approach brings nonprofits, businesses, government agencies, and the public together to tackle complex problems. By doing so, its originators argue, it counters a tendency for social change to focus on the isolated interventions of individual organizations. Kania and Kramer describe five conditions for collective success across diverse stakeholders: "a common agenda, shared measurement systems, mutually reinforcing activities, continuous communication, and backbone support organizations."

The immense popularity of their model has also raised concerns about whether it can truly deliver on its promises. For example, Paul Schmitz, the CEO of Public Allies and author of *Everyone Leads: Building Leadership from the Community Up*, identifies three important areas where Collective Impact may not realize its full potential:[5]

- Enabling leaders to overcome tendencies to tout their own successes, and to be less than honest about what is not working and why they need help from others at the table.
- Encouraging organizations accustomed to seeing an issue from their own point of view to think in a more integrated way about problems and long-term solutions.
- Engaging community members as active leaders and service providers.

Systems thinking can help facilitators of Collective Impact processes, as well as those using other large-scale change approaches, meet the first two of these challenges directly. Table 2.2 summarizes the benefits of using systems thinking within the Collective Impact process.

TABLE 2.2. SYSTEMS THINKING FOR COLLECTIVE IMPACT

Collective Success Conditions	Benefits of Systems Thinking
Mutually Reinforcing Activities	• Develops trust and vulnerability through insights into unintended consequences • Builds understanding of collective *and* individual impact
Common Agenda	• Shared language for communicating interdependencies, delays, and unintended consequences • Shared understanding of root causes of problem and people's contributions to it • Shared aspiration that accounts for benefits of status quo • Shared systemic theory of change
Shared Measurement	• Values qualitative and quantitative data • Assesses progress differently over multiple time horizons • Looks for both intended and unintended impacts • Tracks performance with respect to explicit and systemic theories of change • Aligns goals and metrics with consciously chosen purpose
Continuous Communication	• Quality and consistency of communication improve with increased personal responsibility, deeper alignment around common agenda, and stronger ability to distinguish short- versus long-term impacts • Establishes need for continuous learning as basis for continuous communication

Let's walk through the collective success conditions the table mentions.

First, systems thinking supports *mutually reinforcing activities*. It builds trust by affirming that everyone is doing the best with what they know at the time. It also builds vulnerability by helping people recognize the negative unintended consequences of their well-intentioned actions on both others *and* themselves.

Understanding these consequences enables people to appreciate the extent of their interdependence while simultaneously making their individual accountability for the current situation more transparent. By surfacing the deeper nature of both their connectedness and individual impacts, systems thinking increases the likelihood that people's actions will in fact be mutually supportive.

Second, systems thinking supports the development of a *common agenda* in four ways:

- It provides a shared language for communication. This language, which is described further in the next two chapters, enables people to better appreciate the ways in which they are connected in often non-obvious ways and how time delays and unintended consequences impact their performance.

- It generates a common understanding of why a problem persists despite people's best efforts to solve it. This insight contrasts with a natural tendency to simplify problems in terms of what others are not doing and solutions in terms of what you already do. When you uncover the root causes of a chronic, complex problem, you also establish a solid basis for identifying high-leverage, systemwide solutions.

- It raises the distinction between the desired purpose people espouse and the payoffs of the system as it is currently configured. Without understanding this distinction, stakeholders too easily rally around a shared aspiration while downplaying the strong incentives each of them has to perpetuate the status quo. When people acknowledge both the benefits of change as well as the case for doing business as usual, their vision is grounded in reality, which includes an acknowledgment of what is likely to make change personally difficult. In his "I Have a Dream" speech,

Martin Luther King devoted approximately 70 percent of his time to painting a picture of the difficult reality and only 30 percent to describing the dream.

- It enables stakeholders to create a shared systemic theory of change—a road map of how they will integrate the critical success factors they have identified over time to bridge the gap between what they want and where they are. For example, the community leaders in a school district that served children from both wealthy and poor immigrant families wanted to realize a vision in which all children felt loved and successful. They agreed on more than fifteen critical success factors—including shared measures and the identification and endorsement of a backbone organization. They also agreed on a way to integrate these factors into a coherent strategy to bridge the success gap. (For the specifics of their plan, see chapter 11.)

Third, systems thinking informs *shared measurement*. As Schmitz and others point out, it is easy to be seduced by short-term data and readily measured outcomes even though they might not be indicative of long-term gains. By contrast, systems thinking focuses on both qualitative and quantitative data, assesses progress differently over multiple time horizons, looks for both intended and unintended consequences, and tracks performance in relation to explicit and systemic theories of change. Selecting appropriate indicators also involves aligning goals and metrics with the consciously chosen (rather than de facto) system purpose. For example, ending homelessness calls for metrics that reduce the use of shelter beds over time in favor of permanent housing. By contrast, conventional metrics support using available funds to make more and more shelter beds available to help people cope with homelessness.

Fourth, systems thinking increases the quality and consistency of people's *continuous communication*, because people take responsibility for the impacts of their actions on themselves and others, are aligned around a common agenda as modified above, and understand how to interpret short-term results in a long-term context. In addition, systems thinking emphasizes the need for continuous learning as the basis for continuous communication.

Closing the Loop

- Systems thinking helps people meet four challenges of change: It increases their motivation to change, catalyzes collaboration, enables focus, and stimulates continuous learning.
- Use systems thinking for chronic, complex problems where diverse stakeholders find it difficult to align their efforts despite shared intentions.
- Systems thinking can be used within different change management models. For example, by helping people become more vulnerable and see the big picture, systems thinking supports four conditions for Collective Impact: developing mutually reinforcing activities, building a common agenda, determining shared measurement, and nurturing continuous communication.

CHAPTER 3

Telling Systems Stories

In November 2006, The After Prison Initiative (TAPI), a program of the US Justice Fund of the Open Society Institute (OSI), convened a three-day retreat in Albuquerque, New Mexico, to accelerate progress on ending mass incarceration and harsh punishment in the United States.[1] Aptly named Where Are We Going?, the retreat brought together one hundred progressive leaders—activists, academics, researchers, policy analysts, and lawyers—to clarify what else could be done to facilitate successful reentry of people after incarceration and redress the underlying economic, social, and political conditions and policies that contribute to making the US the world's largest incarcerator among developed nations.

To give you an idea of the scope of the problem, the United States has 2.5 million people behind bars today—versus 200,000 in the 1970s—and approximately 650,000 return home each year. The meeting was grounded in a recognition of how the US criminal justice system—from the beginning and at an accelerated pace since the 1970s—is determined by race, and how society, in the words of Berkeley law professor Jonathan Simon, is increasingly "governed by crime."[2] Most of the participants at the retreat were Soros Justice Fellows or OSI grantees who competed for OSI funding at the same time that they shared a commitment to criminal justice reform.

The challenge presented by this and many similar retreats was that the diverse stakeholders required to solve a chronic, complex problem often do not appreciate the many and often non-obvious ways in which their work is connected. Taking this challenge into account, the goals of the meeting were to:

- Develop a shared understanding of why US incarceration rates and rates at which people return to prison are so high.

- End over-incarceration; create new opportunities for and remove barriers to successful reentry of formerly incarcerated people.
- Strengthen working relationships and collaborations among the advocates.
- Deepen awareness of the interdependencies (both reinforcing and potentially conflicting) among their diverse efforts.
- Identify new ways to strengthen civil society institutions and promote civic and political inclusion.

Perhaps the most radical new tool introduced at the retreat was systems thinking. Working under a grant supported by OSI, the organizers of the retreat, Joe Laur and Sara Schley of Seed Systems, recognized that tackling the same problems with the same mind-set and strategies often produces the same, largely unsuccessful, results. They believed that systems thinking might help people in the field get "unstuck," better understand their theory of change, and devise new strategies and ways of collaborating.

Joe and Sara asked me to introduce systems thinking and systems mapping to help participants create a shared story of why mass incarceration and high recidivism rates persisted, as well as to identify what more they could do to reduce these rates. This picture needed to include the contributions of all participants to the solution, an explanation of why their independent efforts fell short, and insights into what they could do more effectively given limited resources and an urgent need for change.

Storytelling for Social Change

Telling stories is a powerful way to make sense of our own experience and of the world around us. Stories shape our identity, communicate who we are and what is important to us, and move others to act. They are a primary way of distilling and coding information in memorable form. Leaders use them to inspire others. Peace builders recognize narrative as a key source of conflict (people interpret historical facts in very different and incompatible ways), and they work to help disputants both appreciate each other's narratives and modify their own. Therapists use storytelling to help people heal from trauma by supporting them to shape a new and more constructive narrative based on past experience.

Likewise, people committed to social change often share a similar story of what they are trying to accomplish and the challenges they face. Three key elements of this story are:

- The world, in the words of Martin Buber, "stands in need of us," and we are called to contribute our gifts and resources to support those less fortunate than ourselves.
- We are not making the impact we want despite our best intentions.
- The major obstacles to our success are limited resources and the behavior of others in the system.

While the first two aspects of this story are helpful and move people to act in positive ways, the belief that the primary causes of problems are beyond their control holds people back from being as productive as they could be. By attributing shortfalls to limited resources and assuming that others need to be the ones to change, people tend to minimize the impacts of their own intentions, thinking, and actions on their effectiveness.[3] Moreover, because many of the stakeholders compete for limited funds, in this case from The After Prison Initiative, they naturally promote their own successes, downplay their failures, and sometimes may be reluctant to collaborate.

In order to optimize the performance of the entire system, people need to shift from trying to optimize their part of the system to improving relationships among its constituent parts. In the case of US criminal justice, the broader system includes how crime is currently fought, the negative unintended consequences of this system structure, and reformers' efforts to mitigate these consequences and redesign the structure. People need to:

- Understand how focusing on their part of the system—the grantees' reform work in this example—not only supports but might also limit the effectiveness of the whole system.
- Appreciate the non-obvious as well as obvious ways in which they are connected to one another as reformers and to others in the system.
- Recognize the unintended impacts of their intentions, thinking, and actions on both others and themselves.
- Apply this increased self-awareness to shifting how they relate to others in the system.

Even if people's contributions to an existing situation are not obvious, it is important, in the words of Jesse Jackson, that they tell themselves, "We might not be responsible for being down, but we are responsible for getting up." In other words, empowering themselves through greater self-awareness is the first step in changing their reality.

Systems thinking can help people tell a new and more productive story. It honors their individual efforts and surfaces the limitations of these efforts. It distinguishes the short- and long-term impacts of their actions. It aligns their diverse views and stories into a bigger picture where individual contributors can see their part in relation to the whole. Seeing the big picture and their role in it, people are more motivated and able to work together to redesign the whole.

Shaping a Systems Story

In order to tell a systems story, people need to make three shifts:

- From seeing just their part of the system to seeing more of the whole system—including why and how it currently operates as well as what is being done to change it.
- From hoping that others will change to seeing how they can first change themselves.
- From focusing on individual events (crises, fires) to understanding and redesigning the deeper system structures that give rise to these events.

SEEING THE BIG PICTURE

The ancient Sufi story of the blind men and the elephant illustrates the challenge of enabling diverse stakeholders to see the big picture (see figure 3.1). Each party touches a different part of the elephant and tends to assume that what they experience is *the* elephant instead of just one part of a more complex reality. Moreover, they tend to see reality in terms of what they are doing well, are rewarded for doing, and could do better if they had more resources. On the other hand, people either fail to appreciate or question the value of others' contributions. In addition, they often do not have the tools to see a more complex world and understand

FIGURE 3.1 THE BLIND MEN AND THE ELEPHANT. Everyone sees part of a more complex reality and tends to assume that what they see is the whole picture. Sam Gross/The Cartoon Bank

how their intentions, thinking, and actions interact with those of other stakeholders.

In the TAPI case, participants naturally began by seeing solutions to the problem of over-incarceration and failed reentry through their respective specialties. Some focused on sentencing reform to reduce the length of sentences and time served, or the institutional work of resettlement and supportive services, or reorienting parole and probation policies. Others focused on challenging the prison lobby that benefits from current penal laws, or reducing the resistance of public officials to more effective and innovative approaches to reentry. Still others focused on convincing elected officials that tough-on-crime laws make for good politics but bad policy. They entered the group through their own silos. The challenge was to help them expand their perceptions by appreciating how their success depended on the success of all the other stakeholders (including those *not* present at the meeting), and then motivating them to collaborate more effectively with one another (again including those not in the room) to improve public safety in cost-effective and sustainable ways.

The first step was to create a strong and safe container for people to share their different perspectives. This is what I call convening people systemically, and what Marvin Weisbord originally called "getting the whole system in the room."[4] In this case the system was represented in person by those committed to criminal justice reform, while the perspectives of tough-on-crime advocates were depicted on the systems map that included their policies, assumptions, and actions. The facilitators, Joe and Sara, built a container for the retreat participants by building diverse ways of communicating into the agenda, including: expert presentations, panels around specific issues, reports on innovations being tested by several participants, dialogues, a World Café (see more on this and other convening methodologies in chapter 5), and systems mapping.

They incorporated systems mapping because they recognized that convening people systemically is necessary but not always sufficient to mobilizing collaboration. This is true for several reasons:

- Even when people share common values and goals, as those in the TAPI meeting did, they tend to assume that the best way to optimize the system is to optimize their individual part. This assumption is often reinforced by metrics and rewards that encourage people to do what they are already doing.

- By contrast, participants might either fail to appreciate or actually blame (however covertly) others in the room for their inability to be even more effective.

- Some stakeholders are not included in such gatherings because they do not appear to share the same aspiration, are viewed as the source of the problem, and/or are more difficult to access by the conveners. In this case affirming a united front among the participants can mislead them into thinking that they are doing the best they can and others not in the room are to blame. While many TAPI participants were engaged in collaborative efforts with those not present at the meeting, it was important to reaffirm this strategy and avoid the risk of attributing breakdowns in the system solely to other stakeholders.

By contrast, one of the premises of systems thinking as described in chapter 1 is that the best way to optimize the system is to improve the relationships among its parts, not to optimize each part separately. This includes those present in a particular gathering and those who do not participate, those who support change and those who resist it. Helping people who are convening systemically to also *think* systemically enables them to consider collaborating with all stakeholders as a first, though not necessarily the only, option. A systems map enables individual stakeholder groups to see how they contribute to the performance of the system as a whole, both positively and negatively.

For TAPI participants, one of the key insights from the systems map (which is detailed in chapter 7) was that the underlying concern of the public and its elected representatives had more to do with the fear of being victimized by crime and racism than actual crime levels themselves. Although crime levels have actually declined since 1991 by approximately 25 percent, people's fears of being victimized by violent crime continue to rise—as does the perception that crimes are more likely to be committed by people of color, which in turn causes race-associated fear to rise. Even though the criminal justice system consumes enormous tax dollars, public officials who promote mass incarceration often fan fear deliberately to win votes or do so unwittingly by resisting efforts to ameliorate this fear. For example, they resist innovative approaches to resettling formerly incarcerated people (approaches that could reduce recidivism) and fail to distinguish technical from substantive parole violations out of their own fear of appearing soft on crime. This insight led the TAPI participants to think of new ways of collaborating with one another as well as extending themselves to reduce the fears of well-intentioned public officials and concerned citizens who were not at the meeting.

INCREASING SELF-AWARENESS AND PERSONAL RESPONSIBILITY

The natural tendency to view one's own contributions favorably in relation to those of others is intensified by competition. People with a shared aspiration often compete for resources, which increases their reluctance to either acknowledge their own shortcomings or value the contributions of others.

By contrast, a systems story uncovers how people contribute, albeit unwittingly, to their own problems despite their best intentions. Raising self-awareness in this way actually increases their abilities to be more effective. Rather than depending on others to change in order to be successful, they discover that the greatest leverage they have in a system begins with changing themselves. They learn to recognize that taking responsibility for their own intentions, thinking, and behavior gives them more power to create what they want.

Some TAPI participants became more motivated to initiate collaborations with others in the room when they understood the key ways in which they were interconnected. Several also recognized that framing criminal justice reform as a way to help elected officials generate votes by reducing prison costs and recidivism could benefit the reform movement.

UNDERSTANDING THE DEEPER SYSTEM STRUCTURE

One tool for developing an initial picture of "the elephant" (that is, any complex system) is known as the iceberg metaphor. The iceberg is a simple way of distinguishing problem symptoms from underlying or root causes. As shown in figure 3.2, it distinguishes three levels of insight—each of which is informed by a specific question and prompts a certain type of action or response.

More specifically, the iceberg distinguishes the *events* level (what we see most easily) from the *pattern of behavior or trend* that links many events over time, and then goes deeper to expose the underlying *systems structure*—the hidden 90 percent of the iceberg that causes the most damage because it shapes the trends and events. Systems structure includes tangible elements such as the pressures, policies, and power dynamics that shape performance. It also includes intangible forces such as perceptions (what people believe or assume to be true about the system) and purpose (the actual versus espoused intentions that drive people's behavior). The deeper people's level of insight, the greater their opportunity to change the way the system behaves.

People often focus their attention and spend most of their time on responding to individual *events*. They want to know what is happening so that they can react quickly to the crisis at hand. For example, people who support (and oppose) criminal justice reform look at news reports on the latest crime statistics, the number of people recently returned to prison because of repeat offenses or technical parole violations, new legislation,

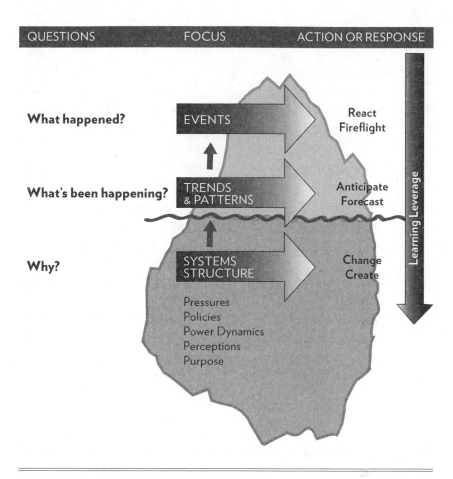

FIGURE 3.2 THE ICEBERG. The iceberg helps you to begin to distinguish a problem's symptoms from its root causes. Innovation Associates Organizational Learning

and costs of the prison system. How people respond to a crisis can have an enormous impact on their effectiveness. Since 95 percent of people sent to prison are eventually released, and many of them are unprepared or unable to resettle productively, get-tough prison sentences often increase recidivism—further destabilizing communities and making them less safe. Moreover, the costs incurred in maintaining the system divert funds that might otherwise be available to strengthen the disadvantaged communities from which a disproportionate number of residents are incarcerated.

Sometimes people step back from individual events long enough to recognize ongoing *trends or patterns*. They ask what has been happening over time and try to anticipate the future based on the past. Trends can often be surprising and disturbing. For example, TAPI participants noted that incarceration levels continued to rise by an estimated 60 percent since crime levels reached their peak in 1991, *despite* a reduction of 25 percent in crime during the same period (see figure 3.3). This led them to conclude that fear, as well as racism, drives current criminal justice policies more than the level of crime itself. Some criminologists believe that no more than 25 percent of crime reduction is attributable to incarceration.[5] Others argue that the same trend data prove the beneficial impact of incarceration on reducing crime, which points to the importance of perceptions or mental models as another aspect of systems structure to be explored below.[6]

The root causes of a chronic, complex problem can be found in its underlying *systems structure*—the many circular, interdependent, and sometimes time-delayed relationships among its parts. The structure includes both easily observable elements—such as current pressures, policies, and power dynamics—and less obvious factors such as perceptions and purposes (goals or intentions) that influence how the more tangible elements affect behavior.

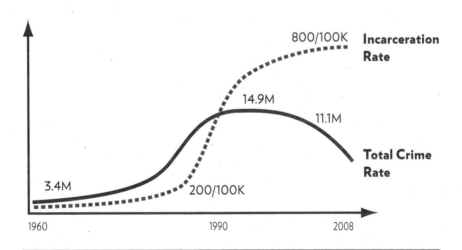

FIGURE 3.3 US CRIME VERSUS INCARCERATION RATES. The growing gap between an increasing incarceration rate and decreasing crime rate raises serious questions about the relationship between the two. Crime rate trend adapted from DisasterCenter.com. Incarceration rate trend adapted from The Hamilton Project, Brookings Institution.

The Elements of Systems Structure

People communicate with one another through language and often through the stories they tell. Michael Goodman, one of the pioneers of the approach used in this book and a longtime colleague of mine, explains that systems thinking can be thought of as a *language*—a visual language that helps us understand and talk about the world in a way that is different from our daily language. The metaphor of language is important because language shapes our perceptions, and hence our behavior. The root of the magical incantation *abracadabra* relates the powers of speech and action, as it comes from either the Aramaic "I will create as I speak" or the Hebrew "It came to pass as it was spoken."[7] In either case, systems thinking is a language that more accurately explains complexity than our everyday language and thus enables us to work more effectively with social systems.

The most basic elements of this language are nouns, verbs, and adverbs (time delays). In addition, when we look more deeply into social systems, we discover that there are certain *plot lines* that appear across a wide variety of issues (whether in education, criminal justice, or homelessness) and at multiple levels of a system (for example, in homes, organizations, or communities).

The most basic plot lines are stories of amplification (called reinforcing feedback) and correction (called balancing feedback). These combine into more complex yet highly recognizable archetypal stories because they are so embedded in the human experience. Knowing the basic stories and systems archetypes gives us initial insights into many chronic, complex problems. Developing a richer and more comprehensive understanding often comes from modifying and combining archetypes—which is similar to illuminating the variations on plots and multiple interacting plots in a historical or fictional story.

Finally, we will look at the bottom of the iceberg to uncover what are described in complexity theory as *attractors*, the pulls that shape and stabilize a system's behavior around a limited number of possible states. These deep structures are the beliefs or assumptions that people in social systems try to validate, and the underlying intentions or purposes they seek to realize. Depending on your assessment of the system's current performance, they can be viewed as either positive or negative. Attractors are the underlying drivers of both system equilibrium and its resistance to change.

BASIC LANGUAGE OF SYSTEMS THINKING

Nouns

The nouns of systems thinking are variables, those forces or pressures at play in the system. Variables "vary" over time; they can increase, decrease, or oscillate. Variables can be qualitative or quantitative and are readily framed as "Levels of ___." Common variables that Michael Goodman and I have identified include what people value (such as the level of expectation or goal), demands on the system (such as the level of need or pressure), resources to meet these demands (such as the level of investment or skills), and actual results (such as the level of performance or effectiveness). They also include perceptual factors that express how people feel and think (such as the level of fear or aversion to risk).

Since variables are the basis for systems stories, defining them is a key task.[8] Significant insights can emerge from clarifying what they are—and what they are not. For example, a breakthrough insight for the TAPI participants was that the fear of being victimized by crime can drive behavior in the criminal justice system more than the level of crime itself. In a very different situation—the effort to rebuild civil society in Burundi after its 1990–94 civil war—NGOs that developed a systems analysis of the conflict determined that the driving factor in the war was not the resources of the Tutsis versus those of the Hutus, as they originally thought, but the power of the elite versus that of the majority. They determined this by recognizing that, when Hutus wrested power from the Tutsis, Hutu leaders became the new elite. In other words, Hutu leaders displayed the same tendency to accumulate resources at the expense of the majority of the population, just as Tutsi leaders had previously fought to retain their power. This insight led them to recognize the importance of another factor, ethnic manipulation, used by elites of both groups to gain and retain power at the expense of their constituents.[9]

Some of the other key variables in the TAPI case were: number of people released from prison, problems with resettlement, technical parole violations, sectors benefiting from the current system, cost of prisons, and (lack of) money available for resettlement. Other qualitative factors included fear for personal safety, political risks, and political resistance to innovation.

Verbs

The fundamental action described in systems thinking is that an increase in one variable causes an increase or decrease in one or more other variables. This action is described pictorially as follows:

$$A \longrightarrow B$$

When a change in **A** causes a similar change in **B** (for instance, an increase in **A** causes an increase in **B**, or a decrease in **A** causes a decrease in **B**), we can put an **s** for "similar" at the end of the arrow.

$$A \overset{s}{\longrightarrow} B$$

Alternatively, if a change in **A** causes an opposite change in **B** (an increase in **A** causes a decrease in **B** or vice versa), we can put an **o** for "opposite" at the end of the arrow.[10]

$$A \overset{o}{\longrightarrow} B$$

While this nomenclature is helpful in building the story, we normally leave it out of the final pictures and instead explain the causal directions verbally on a systems map using descriptive words. This helps people unfamiliar with systems thinking to understand the diagrams.

Time Delay

How long it takes for a change in **A** to cause a change in **B** is a critical factor in systems thinking. This is because, as noted in chapter 1, the short- and long-term impacts of the same action are often reversed. In other words, short-term improvements can produce long-term consequences that neutralize or undermine more immediate gains. Conversely, we often need to invest time, money, and effort in the short run to achieve benefits that are sustainable over time. Time delays are depicted as follows:

$$A \longrightarrow\!\!+\!\!+\!\!\longrightarrow B$$

Michael Goodman and I have identified at least four types of delays in complex social systems. These are the times between:

- The change in a condition and our awareness that the condition has changed.
- Our awareness that the condition has changed and our decision to act.
- The decision to act and the act of implementation.
- Implementation and a corresponding change in the condition.

For example, a current and increasingly serious example is climate change. Although carbon dioxide levels in the atmosphere have increased by more than 45 percent in the past two hundred years, it is only recently that most people have been made aware of the danger of these increases through turbulent weather patterns and rising sea levels. Moreover, because of our dependence on energy-intensive lifestyles and carbon fuels, it has been difficult to mobilize the political will to commit to new energy policies. Assuming we can now make hard decisions, it will still take many years to shift how we conserve energy and manufacture it from environmentally neutral sources. Once we implement these changes, it will take additional time to reduce carbon dioxide levels to necessary levels, though it may already be too late to reverse some changes such as rising sea levels from melting icebergs.

Going back to the TAPI example, there are at least four significant time delays related to the penal system and criminal justice reform:

- The time between when people go to prison and are released— that is, the length of sentences and time spent in prison. Because many sentences have become harsher, it can take many years before people reenter society. The 95 percent of prisoners who are eventually released often face serious barriers to reentry, created in part by the very length of their confinement.

- The delay between the public's fear of crime and their understanding that crime has in fact declined.

- A delay between the number of people incarcerated and concerns about the costs of the penal system. In the years since the TAPI retreat took place, these costs have become even more of a strain on public budgets, reaching an all-time high of eighty-five billion dollars a year, and motivating officials to seriously consider reforms to incarceration.

- The delay between recognizing the costly limitations of mass incarceration and actually shifting funds to the more promising investment of strengthening community institutions—such as education, health care, and employment—that create safer, more prosperous communities.

Because of the pressure to show immediate results—whether self-generated or created by such factors as public opinion, budget cycles, investor expectations, and voting cycles—it can be difficult for policy makers to respect and work with time delays. Leaders can respond more effectively to this pressure when they learn to distinguish *quick fixes* from *short-term small successes*. Quick fixes are solutions that produce short-run benefits, which are typically neutralized or eroded by longer-run consequences of the same actions. Short-term small successes are improvements that are planned from the beginning with the long term in mind and are vital to encouraging persistence and maintaining momentum. This distinction will become clearer when we look at leverage points and strategic planning in greater detail, but these basics will help as we further explore systems plots.

Closing the Loop

- When faced with a complex problem that persists despite their best efforts to solve it, people tend to blame limited resources as well as promote their own successes, downplay their failures, and view others in the system competitively.
- Systems thinking helps people tell a new and more productive story that honors their individual efforts, surfaces the limitations of these efforts, and supports them to see the big picture and collaborate more willingly on behalf of the whole.
- The iceberg metaphor enables people to distinguish between more obvious events and trends, and the underlying systems structure that shapes them.
- Systems structure describes key factors in the system and how they affect one another in often non-obvious ways over time.

—CHAPTER 4—

Deciphering the
Plots of Systems Stories

I love murder mysteries, first made famous by British writers such as Agatha Christie and a staple of popular TV shows like the long-running *CSI*. The essential question they pose is "Who done it?" and the reader/viewer is kept in suspense until the very end in the hope of answering that question. Systems stories are driven by a different question: "Why are people unable to solve a chronic, complex problem or achieve a meaningful goal—often despite their best efforts?" In order to answer this question, it helps to recognize discernible plot lines that tend to shape the behavior of people in social systems.

Many of these plots share a similar and challenging characteristic. Social systems are not only surprising but also, in the words of systems thinker Donella Meadows, "perverse."[1] I think of them as seductive in that they tend to lead people to do exactly the wrong thing for all the right reasons.

Because these plots are so common, they are called systems archetypes. The better people understand them, the less likely they are to become victimized by them. People can learn to anticipate and prevent these stories from seducing them into doing the wrong thing. Alternatively, if people do become trapped, they can follow equally recognizable paths (known as leverage points) to extricate themselves.

Basic Plot Lines

Several years ago, a participant in a systems thinking workshop analyzed a problem he had tried to solve for a long time. He said, "And to think that I have been going around in circles on this issue for years." At that moment I realized that the problem was not so much that he had been "going around in circles," but that he was *unaware* that he was doing so. The solutions he had tried previously were obvious and effective in the short term. However, they had created unintended consequences in the long term that made matters worse. Moreover, when the problem recurred, he failed to see how his own solutions contributed to it. Seeing the circles that he was not only embedded in but also helped create freed him to break out of them and identify a more productive path forward. We go around in circles of our own making without realizing it.

Since systems plots unfold in circles, our goal is to uncover the existing ones so that we can create new and more effective stories. While the emphasis in this chapter is on describing the dynamics—not shifting them, a topic that is more fully addressed in chapter 10—it is important to realize that the act of recognizing the circles you are caught in is the first step toward altering them. *Increasing self-awareness is an intervention in and of itself, and the precursor to making any other changes.*

Reinforcing and balancing feedback are the two basic circular structures that describe how systems evolve over time. More complex dynamics result from combinations of these two feedback structures.

REINFORCING FEEDBACK: THE STORY OF AMPLIFICATION

Reinforcing feedback is the basis for what we know as virtuous and vicious cycles. It explains the development of both engines of growth or flywheels as well as spiraling deterioration. For example, Jim Collins has applied the flywheel concept he introduced in his book *Good to Great* to suggest how social sector organizations can develop their own engines of success.[2] He believes that success in the social sector hinges on the ability to grow organizations (not just programs) by building a brand that attracts support, which yields demonstrable results and in turn strengthens the brand. Collins also points out that the same reinforcing dynamic can produce the opposite effect, as when an organization that performs poorly weakens its brand reputation, which makes it more difficult to attract resources and drives results down even further.

The unstable nature of reinforcing feedback is painfully evident in boom-and-bust cycles such as the housing bubble that set off the 2008 economic crisis. In this case, unsafe subprime mortgage lending practices fueled increased housing prices and more lending—until the bad mortgages could no longer be spread farther and the housing market collapsed.[3]

Reinforcing dynamics also appear in self-fulfilling prophecies. For example, the Pygmalion effect explains how one party's expectations (in this case, a teacher's) lead another party (a student) to behave in ways that reinforce these expectations. This dynamic tends to encourage the performance of well-behaved girls and work against active boys and minorities. The Interaction Map developed by Action Design and shown in figure 4.1 describes these interactions in greater detail.

Most people are accustomed to thinking of growth as linear. However, reinforcing feedback describes a more common process in social and economic systems—that of exponential growth in which a quantity increases by a constant percentage of the whole in a constant time period. Such

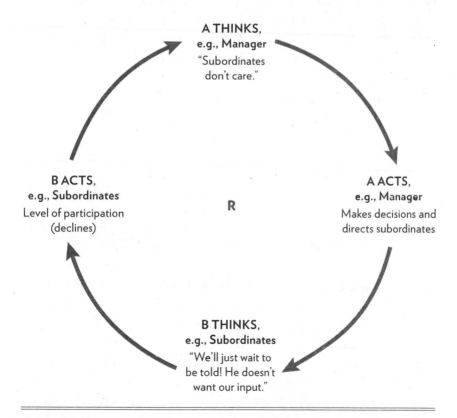

FIGURE 4.1 INTERACTION MAP. How parties A and B think about and behave in relation to each other is mutually reinforcing. Adapted from Action Design

phenomena as increases in savings and population are familiar illustrations of exponential processes. Foundations and entrepreneurs seeking a long-term return on their social investments benefit from cultivating critical mass or tipping points that build sustainable momentum in a social system.[4]

The following French riddle points out several important implications of exponential growth.[5] Imagine a lily pond where the lily plant doubles in size every day, and the pond is totally covered by the lily in thirty days. When is the pond half covered? The answer, which is surprising for many, is day twenty-nine: Half of the pond is covered just one day before the pond is completely blanketed by the lily. How much of the pond is covered in fifteen days? The answer here is 0.0025 percent. In other words, halfway into the month the lily is barely noticeable.

The exponential nature of organic growth has several consequences for social decision makers. First, most people tend to expect to see improvements faster than they are capable of developing. Expecting the system to shift quickly can lead to unrealistic demands for growth that ultimately slow improvement down if not kill it entirely. Alternatively, people can miss or misinterpret small improvements and give up prematurely on supporting a change that takes time to manifest. Figure 4.2 depicts the exponential nature of organic reinforcing growth and contrasts it with the more typical linear assumption people hold about how things *should* grow.

Second, a success engine or flywheel is built not only on the individual factors that contribute to growth, but also on how these factors interact to reinforce one another over time. For example, successful micro-lending programs integrate community involvement, peer support, financial investment, economic results, job creation, and community reinvestment in ever-expanding spirals. An implication for social investors might be that they evaluate grantee plans based on the clarity and soundness of their structural design—how the parts fit together—rather than on the individual elements themselves. We will return to how systems thinking can contribute to articulating such a design or theory of change in chapter 11. For now it can be helpful to notice that one approach to increasing the effectiveness of a theory of change is to explain how parts of the system are intended to interact in both direct and indirect ways over time.

Third, since exponential growth also applies to seemingly trivial problems getting much worse over time, it is important to monitor such problems early on and consider addressing them rapidly instead of hoping they go

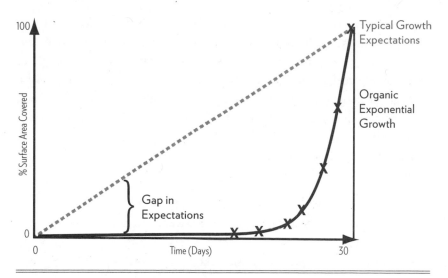

FIGURE 4.2: LESSONS FROM THE LILY POND. People tend to assume that growth occurs more quickly (and linearly) than it actually does. It is important to reduce the resulting gap between expectations and reality. Innovation Associates Organizational Learning and Bridgeway Partners

away. Decades ago, the recognition that small problems fuel bigger ones inspired what's known as the broken windows theory, which suggests that community instability is catalyzed by disorderly conditions.[6] The theory has led police departments around the country to control minor misbehaviors—from littering to vandalism—and maintain a clean environment in order to prevent major crimes from occurring.[7] Critics of the theory argue that petty crime is itself a function of concentrated urban poverty, and that a significant and sustainable reduction in crime levels can only be achieved by improving the quality of life in poor neighborhoods. But either way, the plot is the same: Addressing upstream problems can prevent them from growing exponentially worse.

On the other hand, our failure to address climate change in a timely way represents a serious example of underestimating the severity of a problem by depending on trend data alone. Key decision makers in government and the private sector have resisted recommendations to severely cut carbon dioxide emissions in part because of our dependence on fossil fuels and in part because the problem grew so slowly (as measured by the trend of global temperatures) as to not raise alarms until recently when we are experiencing the effects in real time. An understanding and acknowledgment of the vicious cycles in nature that produced this trend (see appendix A) might have increased political will earlier. Indeed, recent weather patterns and rising sea levels indicate that the curve is likely to have already reached its tipping point as many scientists warned—and our best bet now is to act aggressively to prevent further environmental collapse and figure out peaceful ways of equitably distributing increasingly limited resources.

An understanding of reinforcing feedback can lead foundations, nonprofit leaders, and policy makers to:

- Cultivate the patience to build engines of growth slowly.
- Make decisions based on underlying systems structure instead of trends.
- Break potential vicious cycles quickly.

BALANCING FEEDBACK: THE STORY OF CORRECTION

While the processes of growth and decay might be obvious to many, the dynamics of stability and equilibrium are often dominant and even more difficult to discern. Balancing loops are the driver for improving a social system—we seek to bridge the gap between a current and desired condition—and the key to understanding a system's resistance to change, because the current system is in equilibrium around goals it is already achieving.

We recognize balancing feedback in our daily experience, for example through a thermostat that regulates room temperature at 68°F, or in our own tendencies to sweat or shiver to maintain an internal body temperature of 98.6°F. In contrast with reinforcing feedback loops, which *amplify* an existing condition, balancing feedback seeks to *correct* or reverse a current state by bridging the gap between actual and desired performance. For example, a foundation might fund a mentoring program between older and younger students to improve graduation rates or a counseling program to reduce teen pregnancy. When balancing feedback accomplishes a desired goal, the corrective process often becomes invisible. When we eat enough food or get enough sleep, we tend to take these functions for granted.

By contrast, we are more aware of balancing processes when a system is *not* accomplishing the goal we state for it. In other words, balancing feedback also helps explain why systems do not change despite people's best efforts to improve them. Simple corrective processes fail to function as intended in at least one of three ways.

First, we often stop investing in the solution once a problem appears solved. This act of "taking the pressure off" often leads the problem to recur—much to the frustration of the problem solvers. For example, urban youth crime in Boston was a serious problem in the early 1990s. Political and community leaders banded together to develop numerous coordinated solutions in response—from community policing and neighborhood watches to gang outreach and after-school programs. When youth crime declined as a result, political leaders felt obligated to shift funds to more obviously pressing problems. As a result, they gradually began to cut back on the crime prevention programs that worked so well, and the problem returned.[8]

The second tendency is to fail to appreciate the time required to effect change. For example, a recent success story on curbing teen drinking and substance abuse in one Massachusetts community of forty-six thousand, where adults also exhibited above-average rates of alcohol and drug abuse, described how coordinated improvements had gradually taken hold over a period of eleven years.[9] Such patience and persistence are rare. Normal reactions in the face of time delay are either to become impatient and push for premature results or to give up too quickly.

The third way in which balancing loops can fail to correct an existing situation is when there is lack of agreement on the goals of the system, the current level of performance and what drives it, or both. For example, a report sponsored by the Ball Foundation noted there was no lack of educational innovation in selected US schools and school districts.[10] However, educators seeking to disseminate these innovations on a broader scale were confronted by serious disagreements about both the goals of K–12 education and current performance levels. Some school districts defined their goals in terms of test scores, while others viewed graduation, subsequent employment, or the motivation and capacity for continuous learning as the desired result. Similarly, these school districts measured actual performance differently in terms of test scores, how children performed after graduation, and indicators of creativity and self-directed learning. It is very difficult to define and disseminate a particular strategy when the desired future, system goals, and/or perceptions of current conditions are ambiguous or conflicted.

By understanding ineffective balancing loops, funders, nonprofit leaders, and policy makers can:

- Ensure that effective solutions are reinforced and sustained over time instead of reduced when the pressure decreases.
- Respect time delays by being patient and persistent with social investments.
- Establish a clear and compelling shared vision, joint goals, and a common understanding of current reality before developing strategy. This is the basis for the change model to be introduced in chapter 5.

Figure 4.3 summarizes the core elements of a systems story.

The Plots Thicken

Most complex problems arise from combinations of two or more reinforcing and/or balancing feedback processes. The good news is that we can gain preliminary insight into a wide range of dynamics by becoming familiar with ten of these system archetypes or classic stories. The archetypes are well understood, easily transferable across different system contexts, and often serve as catalysts for discerning even more complex dynamics.[11]This section describes five in greater detail since they illuminate so many problems in social systems, and introduces five more that are helpful to recognize.

FIXES THAT BACKFIRE

Fixes That Backfire is the story of unintended consequences. Figure 4.4 shows the core dynamic of Fixes That Backfire and the pattern of behavior that arises from it. People implement a quick fix to reduce a problem symptom that works in the short run (B1 in figure 4.4); however, the quick fix also creates long-term unintended consequences that exacerbate the problem symptom over time (R2 in figure 4.4). Moreover, people do not recognize these negative consequences as deriving from the quick fix because of the time delay. Therefore, when the symptom returns they incorrectly assume that the solution is to implement *more* of the quick fix. They think, "It worked the first time; we just didn't do enough of it." When they return

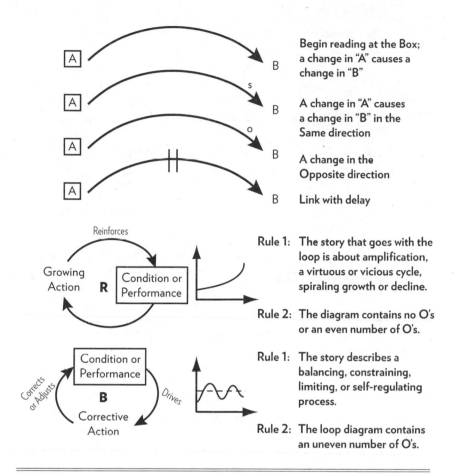

FIGURE 4.3 **CORE ELEMENTS OF A SYSTEMS STORY.** Systems stories are made up of circular cause–effect relationships among variables that change over time. Innovation Associates Organizational Learning

to the quick fix, the cycle repeats itself: short-term gains undermined by long-term negative consequences.

What does a Fix That Backfires look like in practice? Let's return to the TAPI case. The fix of harsh prison sentences reduced crime and the fear of crime in the short run. However, over time prisoners were released, often hardened by their experience or unprepared and legally restricted in their abilities to become productive members of society. On average across the nation, nearly half of formerly incarcerated people succumb to the pressures to commit another crime in the first three years or are sent back to prison

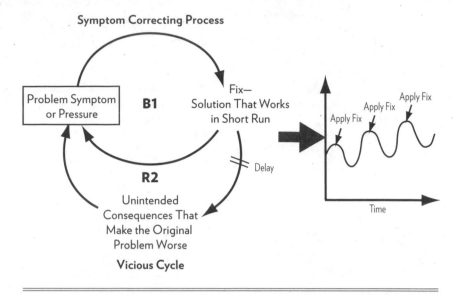

FIGURE 4.4 FIXES THAT BACKFIRE. Fixes That Backfire is the story of a quick fix producing unintended consequences that gradually make a problem symptom worse over time. Innovation Associates Organizational Learning

for parole violations. In a related example, drug busts take criminals off the street and thus reduce drug-related crime in the short run. However, they also remove drugs from circulation, thereby increasing drug prices and requiring addicts to steal more to pay for reduced supplies in the long run.[12]

In health care, as costs of care increase, there is pressure to reduce the length of hospital stays. However, people are often sent home too early and must be readmitted, thereby increasing costs of care even further.

In her book *The Crisis Caravan*, journalist Linda Polman cites the numerous problems created by well-intentioned funding sent by non-profits and wealthier countries to relieve the devastation caused by civil war in poor countries. The relief aid, however well meaning, produces several unintended consequences that exacerbate these humanitarian disasters over time: Fighters in the war become healthier and more able to continue fighting, aid supplies are hijacked by despots or elites seeking to maintain power, and cynical leaders manufacture additional disasters to receive more aid. In addition, the funding provided for relief creates a market for relief organizations that come to compete with one another for more funding.[13]

Another type of relief aid, sending food to people suffering from starvation caused by either human-made or natural disasters, backfires in a different way. The people who benefit the most from food aid are children. Because they survive, they are able to reach childbearing age themselves. Countries receiving food aid then face another spike in population growth and starvation ten to fifteen years after they received the aid.

Cases such as relief and food aid are particularly powerful in raising a poignant and difficult challenge faced by people who want to do good. While there are things people can do to ease others' suffering in the short term, these solutions could make things worse over time. It is incumbent on people who want to help to think through and mitigate the possible unintended consequences of their actions.

Typical keys to overcoming the tendency toward Fixes That Backfire include: questioning the wisdom of the quick fix, identifying an alternative response, or mitigating the negative consequences of the fix if no alternative can be found. Additional possibilities will be covered in chapter 10.

SHIFTING THE BURDEN

In many cases the best way to reduce the likelihood of Fixes That Backfire is to solve the underlying problem that produces the symptoms. People often recognize that a more fundamental solution is desirable, but then wonder why it is so difficult to implement. One of the key reasons is that addressing the root cause of the problem takes longer, is more expensive, and can entail more risk and uncertainty.

This pull between implementing a quick fix and aiming for a more fundamental solution lies at the heart of the so-called philanthropic challenge: Do we fix the problem now or help people over time? In systems terms, depending on the quick fix is known as Shifting the Burden, which produces a similar pattern of behavior as Fixes That Backfire: Intermittent reductions of the problem symptom mask a gradual worsening of the problem. However, there are several important differences:

- In Shifting the Burden people generally know what the more fundamental solution is, but they cannot generate the motivation and investments required to implement it. By contrast, there is no clear fundamental solution to the problem symptom in Fixes That Backfire, and so a quick fix seems like the only possible response.

- In the short run the success of the quick fix, which is the obvious and easier of the two alternatives, creates temporary improvement in the symptom, which in turn undermines people's motivation to implement the more fundamental solution.

- In the long run implementing the quick fix produces unintended consequences that actually undermine people's ability to implement the fundamental solution even if they want to. One common way in which this ability is reduced is that the quick fix consumes resources (people, time, money) that would otherwise be available to solve the problem more permanently.

- As a result people come to depend more and more on the quick fix over time, and invest less and less in the core solution. This growing dependence on the quick fix is also known as addiction. Despite their better judgment, people become addicted to the quick fix.

The systems structure and resulting pattern of behavior are shown in figure 4.5. The top loop (B1) shows the quick fix, while the bottom loop (B2) shows the fundamental solution. B2 is virtual in the sense that it should be activated by the problem symptom but is not; instead the symptom is mitigated by the quick fix to the extent that people do not feel sufficiently motivated to implement a solution that tends to be longer-term and more costly. The combination of B1 and B2 form a vicious cycle that increases use of the quick fix over time while decreasing incentive to use the fundamental solution. The R3 loop on the side shows that increasing use of the quick fix creates side effects that actually decrease the system's ability to implement the fundamental solution over time, thereby exacerbating the problem symptom even further.

The food aid and TAPI cases are examples of both Shifting the Burden and Fixes That Backfire. With respect to food aid, there is a general understanding in the development community that the fundamental solution to starvation is strong local agriculture. However, receiving food aid undermines motivation to develop local infrastructure. In addition the free food

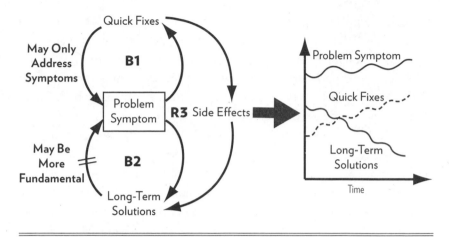

FIGURE 4.5 SHIFTING THE BURDEN. Shifting the Burden is the story of unintended dependency on a quick fix that reduces people's willingness and ability to implement a more fundamental solution. Innovation Associates Organizational Learning

drives down local food prices and makes it difficult for farmers to grow and distribute food profitably, thereby weakening local agriculture even further.

When it comes to criminal justice reform, get-tough prison sentences lead public officials and ordinary citizens to believe that the crime problem has been addressed, thus reducing their motivation to invest in alternative means of solving it. These sentences get offenders off the street, but the burden is shifted when, upon release, formerly incarcerated people are *less* able to do the hard work of resettlement. In addition, the high costs of our current penal system reduce funding for community development and resettlement programs that would reduce crime and the fear of crime in more sustainable ways. Failing to invest sufficiently in community development and resettlement initiatives increases the possibility of crime and its accompanying fears.

In health care, it is common to invest more in treating illness than in preventing it or improving overall health. The long-term consequence of this investment is that there is less money available for influencing the underlying factors that shape good health in the first place.

There are also examples of quick fixes that undermine fundamental solutions to be found in international development. William Easterly, a professor of economics and co-director of the award-winning NYU Development Research Institute, challenges people who are committed to relieving poverty in developing countries to be wary of supporting technocratic solutions implemented by autocrats.[14] He demonstrates that bottom-up development by mostly small actors is much more effective. While top-down technocratic solutions may provide temporary relief for poor people, or at least the appearance of relief, it also takes funds away from the more fundamental solution.[15]

The Shifting the Burden model plays out in the realm of corporate sustainability as well. As John Ehrenfeld, the executive director of the International Society for Industrial Ecology, explains, "Eco-efficiency, or delivering more value for less environmental burden, has been touted as the primary instrument for achieving sustainability. So has socially responsible investing . . . The problem is that none of this espoused benevolence creates true sustainability. At best, it only temporarily slows society's continuing drift toward unsustainability; at worst, it serves as feel-good marketing for products and services that in fact degrade and pollute our environment and fail to meaningfully satisfy the needs of consumers."[16] Ehrenfeld distinguishes between what he sees as the quick fix of supporting more efficient consumption and a fundamental solution that changes the prevailing consumption-driven economic model to one that emphasizes the nonmaterial factors driving quality of life and does not depend on resource-depleting products to create satisfaction.

Peter Buffett, one of the sons of Warren Buffett and chairman of the NoVo Foundation, also calls for redefining the quality of life when he challenges what he calls "philanthropic colonialism."[17] He points out that growing the nonprofit sector is a quick fix to the problem of income inequality because it distracts donors from the deeper work of developing a more humanistic approach to capitalism. Buffet questions the logic of increasing poor people's capacity to consume at the expense of creating a more meaningful experience of prosperity for all. The unintended consequence of depending on the nonprofit sector to solve social problems is that philanthropically minded public- and private-sector leaders can justify what they have earned through a structure that concentrates wealth in their hands by giving some of that wealth back to the poor without challenging the system of inequality itself.

Keys to overcoming the tendency toward Shifting the Burden include: questioning the wisdom of the quick fix, challenging assumptions that discourage investment in the fundamental solution, and establishing a long-term vision that motivates implementation of this solution. Additional possibilities will be covered in chapter 10.

LIMITS TO GROWTH

Limits to Growth is the story of unanticipated constraints (see figure 4.6). Its underlying message is that nothing grows forever. Any engine of growth or success (the R1 loop on the left of the diagram), however effective for a period of time, will inevitably be constrained by external and/or internal factors (that produce the B2 loop on the right side of the diagram). External factors might include the availability of funding, the accessibility of the target population, and the quality of natural resources. Internal constraints might include managerial capability, operational capacity, and an organization's willingness or ability to collaborate with others.

One common example facing most social innovations is the problem of scale-up. Once the innovation is proven, it still faces challenges in expanding its reach to a broader client base. Constraints might come in the form of organizational capacity, funding, and/or ability to create effective partnerships.[18] An example of external constraints is the drain on environmental resources that sustain life as we know it, a problem identified in the pioneering 1972 book aptly titled *Limits to Growth*.[19]

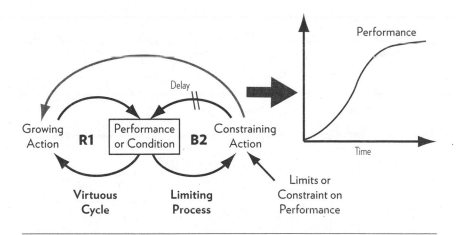

FIGURE 4.6 LIMITS TO GROWTH. Limits to Growth is the story of growth inevitably constrained by limits that must be overcome in order to sustain success. Innovation Associates Organizational Learning

When faced with Limits to Growth, the key steps leaders can take to mitigate the effects of the constraints are to resist the temptation to rely more heavily on the existing growth engine, identify or ideally anticipate the limits, and invest to overcome them using resources provided by the current engine or drawn from outside sources. Chapter 10 describes many strategies for increasing nonprofit capacity and scaling up successful social innovations with the Limits to Growth plot in mind.

SUCCESS TO THE SUCCESSFUL

The tendency to concentrate wealth or success in the hands of the few is itself a common dynamic in social systems (see figure 4.7). In a system with fixed resources, if party A gains an early advantage over party B, A can use that advantage to acquire even more resources (R1, which is a virtuous cycle for A). Meanwhile, party B begins at a disadvantage that grows over time as it becomes less and less able to generate additional resources (R2 is a vicious cycle for B). In other words, opportunity breeds success, success breeds opportunity—and the reverse is also true.

In particular, in his recent book on income inequality *Capital in the 21st Century*, French economist Thomas Piketty points out that accrued benefits to the already wealthy come not just in the form of more goods, but also in the form of capital that makes them even more productive and thus concentrates wealth further and further into their hands.[20] Capital includes savings and inherited wealth that lead to such income-generating investments as stocks, land, higher-quality education, better

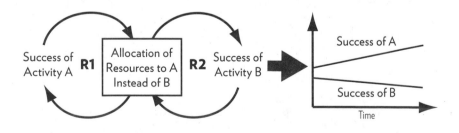

FIGURE 4.7 SUCCESS TO THE SUCCESSFUL. Success to the Successful explains how one party's success and another party's failure can be so closely linked. Innovation Associates Organizational Learning

health care, and access to influential people. By contrast, while money spent by people on acquiring goods may provide more comfort, it does not necessarily increase their access to the factors of production required to create more wealth.[21]

While certain dynamics help the rich get richer, others work directly or indirectly against the poor and especially minorities. This is what experts like Keith Lawrence, co-director of the Aspen Institute Roundtable on Community Change, call structural racism, which he defines as "the nor-malization and legitimization of an array of dynamics—historical, cultural, institutional and interpersonal—that routinely advantage whites while producing cumulative and chronic adverse outcomes for people of color."[22]

Examples of structural racism include gerrymandering and other restrictions imposed on largely minority voters. People who have been in prison, who are also largely black men, face higher hurdles to reenter society when they are released, including a criminal record that often discourages prospective employers. Infants and young children born into poor families get a worse start in life because their parents are often under enormous economic and emotional stress and do not have access to quality health care and preschool services. Recent studies show that the best way to fight inequality is to give these families help early—even before birth.[23]

While it is tempting to associate the Success to the Successful dynamic with capitalism, the tendency exists in most societies: capitalist, communist, and traditional. Sustainable societies moderate it through various redistrib-utive mechanisms that enable all of its members to live in relative balance.

ACCIDENTAL ADVERSARIES

As described in the Collaborating for Iowa's Kids case, Accidental Adversaries is the story of two prospective partners who gradually—and inadvertently—become enemies. As shown in figure 4.8, parties A and B ideally contribute to each other's success through actions they take that benefit the other (outside loop R1). When A for its own reasons is less successful than it wants to be, it independently adopts a solution that improves its own performance (B2). However, its solution unintentionally obstructs B's success. When B is less successful than it wants to be, it adopts a solution to improve its own performance that works for it (B3). However, B's solution unintentionally undermines A's success. The combination of

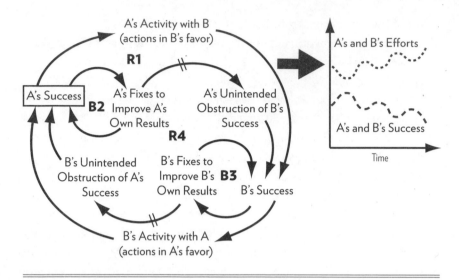

FIGURE 4.8 ACCIDENTAL ADVERSARIES. Accidental Adversaries describes how a promising relationship can unwittingly deteriorate into an adversarial one. Innovation Associates Organizational Learning

independently chosen solutions that inadvertently obstruct each other's performance is a vicious cycle (R4). In essence, parties A and B create Fixes That Backfire on themselves by making life more difficult for the partner that could potentially help them.

The Iowa case is an example of inadvertent conflict between a central organization and its geographic representatives. Stakeholders identified this conflict between the state Department of Education and Area Education Agencies as a system; they likewise identified the conflicts between AEAs as a system and individual AEAs, between individual AEAs and local school districts, and between local school districts and the state Department of Education. The same dynamic created tension between a community college district and the five individual colleges that made up the district. In this case a new college president wanted to centralize functions historically managed by each of the individual colleges in order to increase efficiencies across the district. However, the colleges resisted centralization because they were concerned that losing control of these functions would reduce their ability to customize services for their distinct student bodies.

A very different example is the tension that exists between elected officials and civil servants.[24] Elected officials need the civil servants who work for them to implement initiatives, and civil servants benefit from the political influence provided by these officials. However, shifting political administrations often lead elected officials to implement changes that make it difficult for civil servants to fulfill mission outcomes despite election cycles, and the civil servants in turn seek to maintain mission-critical work.

For example, when William Riley became administrator of the US Environmental Protection Agency (EPA) under President George H. W. Bush, he sought to expand the organization's mission beyond solely regulation to focus more on pollution prevention and conservation. To accomplish this, he sought to integrate EPA programs around a place-based or whole systems approach to environmental outcomes. This required moving away from the siloed structures and programmatic measures of success generated by years of legislative policies.

Then, four years later under the Clinton administration and new EPA leadership of Carol Browner (who to her credit tightened the Clean Air Act's ambient-air-quality standards), the agency reverted to the original siloed structures and programmatic measures of success despite many opportunities to do otherwise with the passage of the Government Performance and Results Act. During that time, however, senior civil servants in the Boston regional office kept Riley's integrated and pollution prevention approach alive through their restructuring of this office to create an ecosystem protection division and pollution and enforcement protection division. They also redesigned the programmatic and individual performance measures by reinventing the performance management system to reflect a place-based, integrated approach to outcomes. They did this out of a strong belief that the public good was best served by focusing on the organization's ends of environmental outcomes, not just the means of permits and enforcement cases, and despite overwhelming resistance from some of their bosses in Washington and Boston as well as some of their peers. In an ideal world, both groups would work together to integrate, improve, or retire fragmented and antiquated laws and policies; establish shared strategic plans and goals that are both long- and short-term; and utilize all existing resources on behalf of the mission and strategic goals.

More generally, the keys to strengthening the partnership between Accidental Adversaries are to clarify the potential benefits of the partnership to both parties, emphasize that the problems caused by both sides have not been intentional, and support both groups to develop solutions to their respective problems in ways that do not undermine the other group.

OTHER SYSTEMS STORIES

Five other plot lines that can be easily recognized across multiple social problems are Drifting Goals, Competing Goals, Escalation, Tragedy of the Commons, and Growth and Underinvestment.

Drifting Goals is the story of an unintentional drift to low performance. It is a special case of Shifting the Burden, where the easiest alternative to implementing a long-term fundamental solution is to lower the goal of the system (thereby reducing the need to make such an extensive investment). For example, in recent years we have come to accept an increasing polarization in US politics, one that has threatened the very functioning of our federal government more than once. We allow this at the expense of effectively challenging the electoral process and the influence of money on political influence. On a more personal note, we have come to tolerate disrespectful language and highly sexualized expression in music and videos available to children (my son is ten) instead of questioning the values that generate them.

Competing Goals comes in two forms: conflicting goals and multiple goals.[25] In the first case, it is impossible to achieve two different goals by taking the same action. In the case of deep-seated conflicts, the goal of defeating one's enemy cannot be accomplished at the same time as the goal of peaceful coexistence. For example, the voices of Israelis and Palestinians who prefer a peaceful two-state solution are gradually drowned out by extremists on both sides who want their neighbors to be eliminated or subjugated instead.[26] By contrast, the problem of multiple goals is one of overload—people trying to accomplish too many goals and therefore being ineffective in achieving any of them.

Escalation is the story of unintended proliferation: The harder you push, the harder your adversary pushes back. Most commonly, escalation describes efforts to dominate or gain revenge on the other party. Arms races and wars are examples of this dynamic, where each party tries to gain advantage over the other by force. Ironically, escalation also explains the "race to victimhood" found in identity-based conflicts where each side seeks to demonstrate that it is the more affected victim of the other's aggression.[27] Psychologist Terrence

Real explains that these tendencies toward aggression and victimization are two sides of the same coin by observing that people tend to "oppress from the victim position" as a way of justifying their aggression.[28]

Tragedy of the Commons is the story of depleting a collective resource that no party feels individually responsible for maintaining.[29]It is most easily recognizable in the destruction of our natural resources—whether overharvesting fisheries and forests, polluting air and water, or exhausting valuable topsoil. A more subtle form within organizations is the tendency of individual departments to place excessive demands on a centralized special resource (such as IT), thereby undermining the effectiveness of that resource over time.

Growth and Underinvestment is the story of self-created limits. By investing insufficiently in a new venture, an organization fails to adequately fund the capacity that would be required to meet growing demand. Because capacity is unable to keep up with emerging demand, the demand itself not only fails to increase but may actually decline. Moreover, the organization then interprets limited demand as a signal that its originally conservative investment was justified, instead of as an indication that sufficient invest-ment in building capacity—not just demand—is the key to long-term growth. Examples of this occur in inadequate funding of new social ventures and restricting investment to expanding an organization's direct services at the expense of developing requisite organizational capacity.

Before closing this section, it is helpful to note one other story line, known as the Bathtub Analogy. This analogy adds the concept of flow to those of stocks (or levels or variables) and feedback relationships introduced so far. The analogy states that the level of water in a bathtub (or carbon dioxide in the atmosphere, homeless people in a city, units of affordable housing in an area, and so on) is governed by the relative flows of water into and out of the tub. If you want to change the level of water in the tub, you have to change the relative rates at which water flows in and drains out, as figure 4.9 explains.

The analogy gained national attention as *National Geographic*'s Big Idea of the Year in 2009.[30] Developed by Professor John Sterman at MIT and described as The Carbon Bathtub, the idea is "simple, really: As long as we pour CO_2 into the atmosphere faster than nature drains it out, the planet warms. And that extra carbon dioxide takes a long time to drain out of the tub." In order to reduce the level of CO_2 in the atmosphere, it is necessary

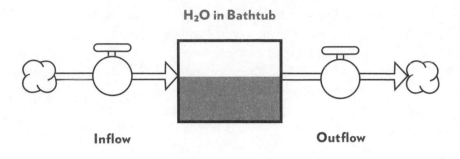

H₂O in Bathtub

Inflow   **Outflow**

FIGURE 4.9 THE BATHTUB ANALOGY. The Bathtub Analogy highlights the importance of understanding stocks and flows when analyzing system behavior. Innovation Associates Organizational Learning

to both reduce CO_2 inflows *and* increase CO_2 outflows, when in fact economic growth and destruction of rain forests are producing the opposite effects. While the analogy seems deceptively obvious, Sterman notes that the tendency to confuse stocks (or levels) with flows is "an important and pervasive problem in human reasoning."

The twelve archetypes (including the foundational ones of reinforcing and balancing feedback) introduced here and summarized in table 4.1, as well as the Bathtub Analogy, form the basis for more complete stories, not their end point. However, these common and recognizable story lines can give people enormously valuable insights into more complex issues. These insights in turn provide the essential self-awareness required to shift less-than-functional dynamics.

The Stories Behind the Story

The dynamics described in the stories above are in turn perpetuated by two other key factors: people's *beliefs and assumptions* about how things should work, and their underlying *intentions* (or purpose). In other words, the system behaves the way it does in part because people are trying to prove that their assumptions are true and to achieve certain goals that they might not even be aware of or acknowledge.

In the case of Collaborating for Iowa's Kids, the fundamental belief held by each organization was that it was doing the best it could to improve K–12 education for children at its level (statewide, regional, or local), and

TABLE 4.1. SUMMARY OF SYSTEMS ARCHETYPES

Virtuous/Vicious Cycles	**Amplification and Reinforcement:** A reinforcing process producing success or disaster.
Balancing Process	**Correction:** We try to reduce the gap.
Fixes That Backfire	**Unintended Consequences:** The long-term negative consequences of a quick fix.
Shifting the Burden	**Unintended Dependency:** The quick fix we become addicted to.
Limits to Growth	**Unanticipated Constraints:** The limiting mechanism on spiraling growth.
Success to the Successful	**Winner Takes All:** Your success produces my failure.
Accidental Adversaries	**Partners Who Become Enemies:** Two parties want to cooperate, but each sees the other undermining its success.
Drifting Goals	**Inadvertent Poor Performance:** Actual and desired performance levels gradually fall.
Competing Goals	**Conflicting or Multiple Commitments:** Trying to satisfy conflicting goals or achieve too many can lead to accomplishing none.
Escalation	**Unintended Proliferation:** The harder you push, the harder the competitor pushes back.
Tragedy of the Commons	**Optimizing Each Part Destroys the Whole:** Everyone takes advantage of a resource that doesn't belong to anybody.
Growth/Underinvestment	**Self-Created Limits:** We push on the growth side and underinvest in the capacity to grow.

Source: Innovation Associates Organizational Learning and Bridgeway Partners

that shortfalls in educational performance were caused by organizations in the system other than itself. The purpose of each organization was to optimize performance across the geographic area for which it was responsible, which it incorrectly assumed would optimize performance for all children throughout the state.

In the case of The After Prison Initiative, advocates of reform believe that rates of incarceration continue to increase despite decreasing crime levels because of structural racism. Others say that high incarceration rates have in fact caused crime levels to drop, although they also note that the marginal benefit of continuing to increase the number of people in prison might not be justified in terms of corresponding decreases in crime.[31] Reformers perceive that the underlying purpose of harsh prison sentences is to marginalize people of color and other minorities because they are different, while many elected officials argue that the purpose of public safety is being achieved by get-tough sentencing.

Being able to recognize all these plot lines in a systems story helps enormously as we enter the next stage—managing change through the four-stage process.

Closing the Loop

- Systems structures can be summarized in terms of recognizable story lines or plots that recur across a wide variety of social issues.
- The key drivers of systems stories are what people assume to be true and their underlying intentions.
- There are several ways to shift these dynamics. The first step in all cases is to become aware of them and one's role in perpetuating them.

PART TWO

THE FOUR-STAGE
CHANGE PROCESS

CHAPTER 5

An Overview of the Four-Stage Change Process

In the summer of 2006, Michael Goodman and I supported a group of community leaders in Calhoun County, Michigan (population one hundred thousand), to develop a ten-year plan to end homelessness.[1] The agreement forged by government officials at the municipal, state, and federal levels—along with business leaders, service providers, and homeless people themselves—came after years of leadership inertia and conflict regarding what needed to be done to solve the problem. Moreover, the plan signaled a paradigmatic shift in how the community viewed the role of temporary shelters and other emergency response services. Rather than see them as part of the solution to homelessness, people came to view these programs as among the key obstacles to ending it.

The plan won state funding, and a new organization led by an executive director and multisector board was formed to steer implementation. Service providers who had previously worked independently and competed for foundation and public moneys came together to work in new ways, as exemplified by their unanimous decision to reallocate HUD funding from one service provider's temporary housing program to a permanent supportive housing program run by another provider. Jennifer Bentley, who chaired the planning process, observed, "I learned the difference between changing a particular system and leading systemic change." In the plan's first six years of operation, from 2007 to 2012, Calhoun County did a remarkable job of securing permanent housing for the homeless, especially in the face of the economic downturn of 2008–09. Homelessness decreased by 14 percent (from 1,658 to 1,419) *despite* a 34 percent increase in unemployment and a 7 percent increase in evictions.[2]

Why was this intervention so successful when many other collective attempts to improve the quality of people's lives fall short? The local foundations and other partners who were involved combined two significant interventions: a proactive community development effort that engaged leaders in the three major sectors along with homeless people themselves, and a systems analysis that enabled all stakeholders to agree on a shared picture of why homelessness persisted and where the leverage lay in ending it. In other words, the approach merged more conventional processes that facilitate *convening* people systemically (such as bringing the whole system into the room) with tools that helped the stakeholders transcend their immediate self-interests by *thinking* systemically as well.

Convening and Thinking Systemically

Part 2 of this book enables you to integrate these processes of convening and thinking systemically into a four-stage change process. Michael and I developed the approach as a result of working with hundreds of executives and change agents who were responsible for leading systemic change and did not know how to use the power of systems thinking to increase their effectiveness. The approach also supports people interested in systems thinking to integrate it into the world of practical action and performance improvement.

Leaders of social change recognize the power of convening multiple and diverse stakeholders—including those representing the nonprofit, public, and private sectors—to achieve breakthroughs around issues that affect all parties. The past twenty-five years have seen many innovations in large group interventions designed to increase communication across the sectors, such as Future Search, Open Space Technology, and the World Café.[3] These interventions are often structured as individual events or a series of events. In addition, new processes such as Collective Impact (reviewed in chapter 2), Theory U, Social Labs, and collaboration partnerships for environmental sustainability have also emerged.[4]

Michael and I have found that systems thinking can enhance many of these approaches by providing what Otto Scharmer, the creator of Theory U, calls a "collective sensing mechanism" that enables all stakeholders to see the bigger picture each of them is part of. This mechanism also helps people appreciate how they not only support but also often unwittingly undermine system performance, thereby empowering them to think and act more effectively.

The Four-Stage Change Process

We built our four-stage change process on the "creative tension" model introduced by Peter Senge in *The Fifth Discipline*.[5] This model proposes that the energy for change is mobilized by establishing a discrepancy between what people want and where they are (see figure 5.1). If people hold to the vision of what they want and are simultaneously clear and candid about where they are, then the tension will tend to resolve in favor of what they want. This principle applies both at the individual level and at the collective level.

Translated to the collective level, when people have a common aspiration—as expressed by a shared vision, mission, and set of values—and a shared understanding of not only where they are now but also *why*—then they establish a creative tension, which they are drawn to resolve in favor of their aspiration. Developing a shared understanding of why the current reality exists is essential to addressing the challenge that stakeholders often

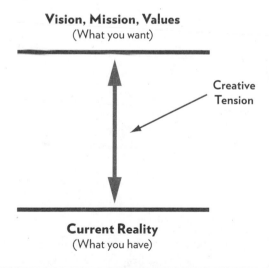

Vision, Mission, Values
(What you want)

Creative
Tension

Current Reality
(What you have)

FIGURE 5.1 ESTABLISHING CREATIVE TENSION. Energy for change is created by the tension between a desired and the actual condition. Innovation Associates Organizational Learning

agree on where they are at the top of the iceberg (for instance, feeling pressure to build another shelter), but they fail to see the underlying systems structure that affects and is affected by all of them (their dependence on temporary shelters as a solution to homelessness).

Developing a shared picture of what people want as well as of reality at a deep level enables stakeholders to experience their responsibility for the whole system instead of just their role. It produces a state of *alignment* where people freely commit, "I'll get my part done, and I'll make sure we all get the whole thing done." For example, they might question a decision to invest in new shelters and promote additional investment in permanent, safe, affordable, and supportive housing instead.

So we expanded the creative tension model into a four-stage change process where stakeholders:

1. Build a foundation for change and affirm their readiness for change.
2. Clarify current reality at all levels of the iceberg and accept their respective responsibilities for creating it.
3. Make an explicit choice in favor of the aspiration they espouse.
4. Begin to bridge the gap by focusing on high-leverage interventions, engaging additional stakeholders, and learning from experience.

The process is summarized in figure 5.2.

STAGE 1

The purpose of Stage 1 is to build a foundation for change. The intended result is to develop collective readiness for change. Stage 1 incorporates three steps:

- Engage key stakeholders. This involves identifying the range of possible stakeholders and designing strategies to engage them individually and collectively.

- Establish common ground by creating initial pictures of what people want to achieve and where they are now. It is useful at this point to develop an initial shared vision of the ideal outcomes and an overview of what is and is not working now.

- Build people's capacities to collaborate with each other. This involves developing people's abilities to think systemically and hold productive conversations around difficult issues, as well as their underlying capacity to take responsibility for current reality.

For example, the Calhoun County project brought together leaders from the public, private, and nonprofit sectors along with homeless people themselves to develop their ten-year plan to end homelessness. They engaged in shared visioning and were introduced to productive conversation and systems thinking tools early on in the process.

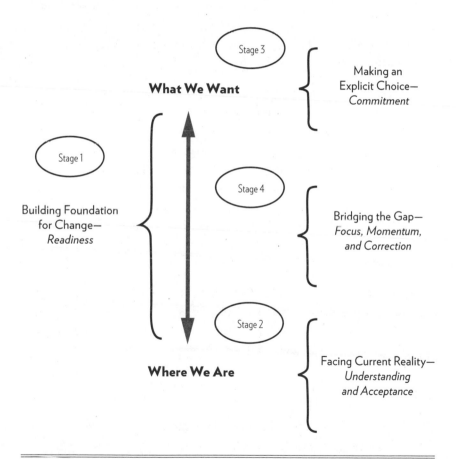

FIGURE 5.2 FOUR STAGES OF LEADING SYSTEMIC CHANGE. This four-stage model provides a clear path for leading systemic change. Bridgeway Partners and Innovation Associates Organizational Learning

STAGE 2

The purpose of Stage 2 is to help people face current reality. The intended results are to build not only a shared understanding of what is happening and why, but also acceptance of people's responsibilities—however unwitting—for creating this reality. While it might seem logical at this point to develop a clearer and richer picture of the ideal future, Michael and I have found that digging more deeply into reality at this stage more accurately reflects many people's desires to understand and be understood for where they are before venturing too far forward. As Otto Scharmer observed based on his work with Ed Schein, professor emeritus of management at MIT, the primary job of leadership is to "enhance the individual and systemic capacity to see, to deeply attend to the reality that people face and enact."[6]

The tasks of Stage 2 are to:

- Identify people to interview about the history of the current situation and clarify what questions to ask.
- Organize and begin to improve the quality of the information.
- Develop a preliminary systems analysis of how different factors interact over time to support or undermine achievement of the vision.
- Engage people in developing their own analysis as much as possible.
- Surface the mental models that influence how people behave.
- Create catalytic conversations that stimulate awareness, acceptance, and alternatives.

For example, Michael and I interviewed fifty leaders from across the three sectors as well as homeless people, developed an initial systems map that we initially vetted with a small design committee, and refined and then shared our analysis with a wider steering committee.

STAGE 3

The purpose of Stage 3 is to help people make an explicit choice in favor of what they really want. The intended result is that they consciously commit to their highest aspirations with full awareness of the costs, not just the benefits, of realizing them. The steps to achieve this result are to help stakeholders:

- Identify the case for the status quo uncovered in Stage 2—the short-term benefits of the current system such as quick fixes that work, for instance, and the immediate gratification that comes from implementing them—and the costs of changing, such as the need to make longer-term investments in effort, time, and money.
- Compare this with the case for change described in Stage 1—the benefits of change and the costs of not changing.
- Create both/and solutions that achieve the benefits of both—or be willing to make hard trade-offs between them.
- Make an explicit choice and bring it to life through a vision that illuminates what people feel called to or deeply wish to create.

A critical turning point in Calhoun County's ten-year planning effort occurred when the stakeholders realized that their current system was perfectly designed to help people *cope* with homelessness; however, the way they were operating actually undermined their avowed purpose to *end* homelessness.

STAGE 4

The purpose of Stage 4 is to help people bridge the gap between what they deeply care about, which they affirm in Stage 3, and where they are now as clarified in Stage 2. This final stage involves identifying leverage points and establishing a process for continuous learning and expanded engagement. Specific tasks are to:

1. Propose and refine high-leverage interventions with community input. This includes:
   a. Increasing people's awareness of how the system functions now.
   b. "Rewiring" causal feedback relationships.
   c. Shifting mental models.
   d. Reinforcing the chosen purpose through updated goals and plans, metrics, incentives, authority structures, and funding.
2. Establish a process for continuous learning and outreach. This covers:

    a. Engaging existing stakeholders on an ongoing basis.

    b. Developing an implementation plan that incorporates demonstration projects as part of a long-term road map.

    c. Refining the data to be gathered based on new goals and metrics.

    d. Evaluating and revising the plan regularly with input from current stakeholders.

    e. Expanding stakeholder involvement by tapping additional resources and scaling up what works.

In the case of Calhoun County, the leverage points people identified became the goals of their ten-year plan, and others have become involved in the implementation. Broader involvement to end homelessness includes engaging stakeholders responsible for economic development, affordable housing, foster care, and the criminal justice system.

While the four stages and supporting tasks are listed numerically, the process itself is not always linear, as we'll see in the chapters ahead, which map out the tasks associated with each step. The lessons from Stage 4, for instance, feedback into a new Stage 1 in an ongoing, circular process. Allowing ample time for the process is critical: In this case, the shortest distance between two points is indeed a circle.

Closing the Loop

- You can increase your ability to lead systemic change by integrating ways of convening multiple, diverse stakeholders with tools that help them think systemically.
- One proven approach for combining the two is a four-stage change process that explicitly harnesses the power of systems thinking.
- The four stages are: building a foundation for change, facing current reality, making an explicit choice, and bridging the gap.
- These stages will be detailed in chapters 6 through 10.

CHAPTER 6

Building a Foundation for Change

Imagine engaging a group of community stakeholders to address an important social issue—such as ending homelessness, strengthening K–12 education, or improving local public health. You would want to identify who to bring together and how, establish common ground among the participants, and support them to collaborate with one another.

Now imagine your first group meeting and being confronted by the fact that people have actually come with two different agendas: their public one to address the issue and their private one to optimize their part of the system. John McGah of the homelessness initiative Give US Your Poor and I developed table 6.1 to distinguish these two agendas for participants in a typical homeless coalition meeting.

How would you address the challenges of different interests and perspectives to build a strong foundation for change? How would you ensure that you invite the right people in the first place, establish common ground, and develop their abilities to work together?

Engage Key Stakeholders

Key stakeholders are people and organizations that affect and are affected by the issue. They include anyone that can make a contribution to the effort, or anyone that can possibly derail it if not on board. Broadly, participants might include nonprofit organizations representing community interests and/or specific populations, government agencies that are charged with developing or implementing social policies, law enforcement, health providers, schools, businesses concerned with the impact of an issue on economic development, the media, and members of the target population. Diversity is key because systems depend on it to innovate.

TABLE 6.1. THE HOMELESS COALITION MEETING

Role	Espoused Purpose	Hidden Priorities
Elected Official	Permanent housing with support services and jobs are important.	This takes a long time and is expensive; the community has other more immediate issues; I need to be reelected shortly.
Business Leader	It's important for everyone to have shelter.	Our primary concern is homeless people downtown who hurt business.
Homeless Shelter Director	Giving people shelter is humane.	The more beds we fill, the more money we get.
Director of Health Care for Homeless	Homeless people need basic health services outside the ER.	We have to compete with other providers for limited funding.
Affordable Housing Advocate	All people need permanent housing first.	We need to attract people who can afford to pay for housing and have less complex needs.
Funder	We are committed to helping homeless people.	Our board wants to show results *now*.
Concerned Citizen	No one should be homeless, and shelters provide a humanitarian solution.	I don't want homeless people living near me; taxes should go to more pressing problems.
Homeless Person	Permanent housing gives me ongoing security.	My community is other homeless people; I don't know if I can make it in mainstream world.

Source: Bridgeway Partners and Give US Your Poor

In order to engage key stakeholders, a convening organization or group such as a foundation or community-wide board needs to clarify who should be actively involved and then develop a strategy for getting them to work together. It helps to include the following core group members:

- Executive sponsors and key decision makers representing the key constituencies who have a deep interest in the issue and opportunity.
- Activists with a personal passion for the issue.

TABLE 6.2. ANALYZING KEY STAKEHOLDERS

Name	Current Support (-3 to +3)	Desired Support (-3 to +3)	Their Motivation	What You Can Do

Source: Innovation Associates Organizational Learning and Bridgeway Partners

- Ultimate beneficiaries who usually have little or no voice in the current system, such as patients, students, homeless people.
- A professional consultant or facilitator.

A stakeholder map is a simple tool to guide the engagement process and expand participation (see table 6.2). For example, in applying the tool to end homelessness, use column 1 (NAME) to identify the groups or individuals who need to be involved because they impact or are impacted by the issue. In column 2 (CURRENT SUPPORT), consider how supportive each stakeholder currently is of creating a new reality, on a scale of -3 to +3. A -3 indicates that they are strongly motivated to block efforts to end homelessness (for whatever reason), a 0 indicates neutrality, and a +3 indicates that they are fully motivated to take the lead in ending homelessness.

In column 3 (DESIRED SUPPORT), write down how you as a convener want each stakeholder to be involved in ending homelessness. For example, you might want to move a group that is currently a -3 (looking to block the effort), -2 (strongly opposed), or -1 (somewhat opposed) to a more neutral 0 position. Or you might want to motivate a currently neutral group to become a +1 (somewhat supportive) or +2 (strongly supportive) contributor. Since it helps to have only one organization or group such as a multisector leadership board in a formal leadership role, identify one stakeholder whom you want to see in the +3 role.

In column 4 (THEIR MOTIVATION), clarify the motivators for each stakeholder to participate in the way you described in column 3. Some motivators are likely to be the same for many stakeholders, while others will be unique to specific groups. If people are resistant to change, clarify in this column the nature of their resistance as technical, political, or cultural. Note in column 5 (WHAT YOU CAN DO) how you intend to engage each stakeholder depending on why they would want to be involved. Some groups might be best engaged initially through individual outreach, while others might be glad to be involved through a community-wide coalition. If people are resistant to change, recognize that you still have several options. You can legitimize and address their concerns directly, influence them through others, engage them at critical phases in the process, or work around them.

The form of collective gathering will vary according to the nature of the issue. For example, in the case of the Collaborating for Iowa's Kids project, a core group of leaders from the state Department of Education and Area Education Agencies convened a larger group of representatives from both organizations, and subsequently invited in representatives from Local Education Agencies (LEAs) as well when it became clear that LEAs needed to be part of developing a new collaborative process.

In efforts to end homelessness, stakeholders are likely to include individuals and organizations that make up the Continuum of Care, the coordinating body defined by the Department of Housing and Urban Development to address homelessness in a geographic region. It is important to think about people who provide not only shelter and housing for the homeless but also related services—such as child welfare, criminal justice, health care, transportation, and education—that impact and are impacted by the problem. Perhaps less obvious but equally critical are businesspeople, because they affect and are affected by the economic health of the community, which in turn impacts homelessness; public-sector officials at the municipal, county, state, and federal levels, because they influence funding streams and policy related to homelessness; and homeless people themselves. It is also important to consider how to involve the media and shape citizen opinions about the issue.

Conveners need to address several challenges in engaging larger groups of stakeholders:

- Onetime events, such as the retreat used in The After Prison Initiative, are less likely to be effective than longer-term processes. Although face-to-face processes are more expensive to manage, especially when people are widely distributed at a national or global level, they can be increasingly complemented by virtual work that's supported by improved communications technology. Technology reduces overall costs, and the combination of face-to-face and virtual collaboration produces the benefits of sustained collective attention.

- Asking people to propose reforms to an existing system can lead them to think that they are not part of the system, and hence not part of the problem. Systems thinking enables people to see how they are part of the problem, which ironically increases their ability to develop effective solutions.

- Reformers often blame powerful stakeholders who represent the status quo and are not part of the redesign process for their own inabilities to effect change. While certain stakeholders do resist change, it is important to realize that there are several ways to work with this resistance, including: legitimizing and addressing their concerns directly, influencing them through others, and engaging them at critical phases in the process. Alternatively, it might be necessary to work around them or use more activist strategies such as political advocacy, active opposition, and legislation to change policy—although these are not the focus of this book.

Establish Common Ground

Establishing common ground involves developing an initial appreciation of why people are coming together, a shared sense of direction, and agreement on some of the key aspects of current reality.

What brought people together in the Collaborating for Iowa's Kids project was a common concern that, despite reform efforts implemented over the past decade, student performance was not increasing in relation to the state's own high standards and relative gains demonstrated by kids in other states and counties. The Open Society Institute convened The After Prison Initiative retreat to clarify why incarceration and recidivism rates remained so high despite participants' extensive efforts to reform the criminal justice system. Leaders in Calhoun County came together to capitalize on an opportunity to receive state funding for developing a ten-year plan to end homelessness in their community.

One useful tool in establishing a common rationale for coming together is to ask people to identify a *focusing question* they want to answer. The focusing question is a way of helping people define the boundaries of a systems analysis. Since everything is ultimately connected to everything else, the question enables them to develop a rich yet manageable level of insight into the root causes of a chronic, complex problem. It asks, "Why, often despite our best efforts, have we been unable to achieve a certain goal or solve a particular problem?" The question "why" is essential because this leads people to uncover root causes; by contrast, "how to" questions mobilize them to implement solutions to problems they often do not fully understand.

The use of a focusing question points to a paradox of systems mapping: *The purpose of systems mapping is to answer a focusing question—not to map an entire system.* Answering a focused question is a bounded objective that yields actionable insights, while mapping an entire system can be an unbounded task that produces confusion and paralysis in the name of comprehensiveness.

Developing a shared sense of direction involves clarifying the mission, vision, and values of the convening group on behalf of the stakeholders they represent. For example, my colleague Kathleen Zurcher helped the convening group for Collaborating for Iowa's Kids to define "the Hallmarks of Our Partnership" and "the Future We Will Create Together."[1] The hallmarks were people's mission and core values in coming together. They described their desired future in terms of both a vision statement and a detailed picture of their end result. Their rich picture answered two questions: "What will we experience in Iowa when this vision is achieved? What difference will it make?"

The members of the 10-Year Planning Committee to End Homelessness in Calhoun County summarized their vision as:

- A comprehensive, integrated implementation plan to reduce homelessness and chronic homelessness in Calhoun County.
- A strong coalition of service providers, homeless individuals, funders, and community leaders with a community-wide commitment to end homelessness.
- A system that fosters collaboration efforts and a team approach to end homelessness.

Several factors influence how much time to spend on visioning in this first stage. Kathy points out that when the quality of relationships among stakeholders is very fragile or people are too overwhelmed by current circumstances to be creative, it can help them to spend more time on cultivating a shared vision before moving to inquire deeply into the way things are. On the other hand, if people are feeling disconnected from what is happening now or frustrated by their inabilities to implement "obvious" solutions, then it makes sense to move to Stage 2 faster.

The final step in developing common ground is to highlight key aspects of current reality in relation to the vision. For example, the community leaders in Calhoun County contrasted their vision with the following observations about the way things are:

- Although we have many agencies working on different important aspects, we need stronger team approaches.
- There is not a lot of public education about homelessness.
- Our current coalition is made up of primarily service providers without the needed community, resident, and monetary support.
- Not everyone is aware of other agency services.

You can use the iceberg tool to highlight current reality at multiple levels:

Level 1: Important events that have triggered people's desire to come together—such as Calhoun County's onetime opportunity to receive state funds to end homelessness.

Level 2: Relative changes in key indicators over time—such as growing incarceration rates despite declining crime rates.

Level 3: Critical pressures, policies, and power dynamics that affect the issue or opportunity—such as the impact of structural racism on efforts to reform the criminal justice system.

Level 4: Underlying assumptions or mental models—such as the assumptions in Calhoun County (and elsewhere) that "people want to be homeless" and "the individual is the problem, not the system."

Creating a common context for collaboration and establishing creative tension through initial statements of a shared direction and contrasting current reality help provide a strong foundation for change.

Build Collaborative Capacity

The last cornerstone of a strong foundation is developing people's abilities to work with one another. Introducing these skills at this stage is important because optimizing the system requires improving the *relationships* among its parts, not optimizing the individual parts as is often assumed and rewarded. Improving the whole also requires that people feel comfortable sharing information that is as timely, accurate, and complete as possible.

One capacity to develop is thinking systemically. Supporting people to use the language of systems thinking increases their abilities to see the bigger picture and speak in ways that take this picture into account. It can be especially helpful at this stage to introduce several of the principles and tools covered in earlier chapters of this book:

- Good intentions are not enough.
- Characteristics of failed solutions.
- Conventional versus systemic thinking.
- The iceberg.
- Reinforcing and balancing feedback.
- Time delays.
- Common systems archetypes.

When people come to understand that they are connected in non-obvious and often counterproductive ways, they begin to appreciate the bigger picture and not just their part of it.

A second capacity is to develop productive conversations around difficult issues. As the metaphor of the blind men and the elephant illuminates, people seeking to work together often have very different views of reality. In addition, the example of the homeless coalition at the beginning of this chapter shows that even people with shared aspirations can have very different secondary agendas. People who want to achieve social change need to learn to engage and bridge differences.

The core skill for productive conversations is to recognize that the world is much more complex than people think. Our assumptions or mental models are at once useful, limited, and capable of becoming more accurate. For example, assuming that "street people prefer to be homeless" might be useful in that it acknowledges that they might have difficulties in adjusting to living in a permanent home. The same assumption is limited in the sense that most chronically homeless people who are given the opportunity to live in permanent housing with support services take advantage of it; in one case 96 percent were still living in the same housing one year later.[2] Hence, a more evidence-based and accurate assumption is that most street people prefer to live in permanent housing if it is safe, is affordable, offers community, and is coupled with counseling services.

The Ladder of Inference (see figure 6.1) is an excellent tool for helping people distinguish what they think from the larger reality around them. It shows how people select certain data out of an almost infinite pool of available data, make assumptions and draw conclusions based on the data they select, make recommendations and take action based on these conclusions, and then look for new data that reinforce their original assumptions.

Another useful tool is what Peter Senge describes as "balancing advocacy and inquiry."[3] Most people are more accustomed to advocating than inquiring, so it often helps to begin with inquiry—the art of asking others how they see the world and then listening to them deeply. As my colleague Bryan Smith told me many years ago, people need to know that you care before they care what you know. Ask others:[4]

- What do you see (the observable data)?
- How do you feel as a result of seeing those data?
- What do you think or tell yourself as a result of those data?
- What do you want?

Then really listen. Otto Scharmer distinguishes four levels of listening: listening for what you already know, listening for what surprises you, listening with empathy for the other's experience, and listening from a deeper source that seems to embrace your truth and theirs.[5]

Once you have established that you care about others' views, you can be a more effective advocate for your own. Since each of us sees part of a more complex world, it is also important that you be able to contribute to people's understanding by advocating your view. In order for your advocacy to be heard and used most effectively, it helps to learn to advocate so as to both share what you know *and* invite others to comment on and potentially enhance your knowledge. Effective advocacy involves understanding and making transparent your own Ladder of Inference so that others can add to and improve upon the data, reasoning, and conclusions you have drawn.

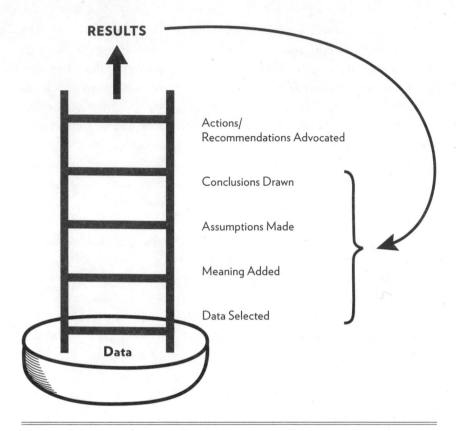

FIGURE 6.1 THE LADDER OF INFERENCE. The Ladder of Inference shows how people unconsciously jump from data to conclusions. Based on the work of Chris Argyris and Don Schon

By balancing advocacy and inquiry, you create not only a more accurate picture of what is and how to work with it, but also more support from others for taking effective action.

The third capacity is to cultivate a viewpoint of responsibility. Both thinking systemically and holding productive conversations develop a deeper capacity to understand how you are responsible for a situation as it currently exists, not just for solving it. Taking responsibility for the problem as it exists does not mean blaming yourself for it. It means empowering yourself. You see how your intentions, assumptions, and actions have unintentionally contributed to the problem you want to solve. It is ultimately easier to change how you think and behave than to try to change others in the system.

Even if you are not responsible for the problem, you can use this perspective to ask how your efforts to solve the problem might unintentionally be undermining your ability to do so. For example, if your intention is to convince others that they are wrong and must be the ones to change, then you can activate or embed an adversarial relationship, which is even more difficult to resolve. You create unnecessary opposition when, in the words of master therapist Terrence Real, you "oppress from the victim position."[6] It helps to remember that respect, inquiry, and empathy are often the best keys to use first to open the door of social change.

Closing the Loop

- Begin building your foundation for change by identifying and involving important stakeholders.
- Recognize that there are multiple ways to engage people who resist change—not just those who support it.
- Establish common ground by identifying a common reason for coming together, developing a shared direction, and sketching an initial picture of current reality.
- Build people's capacities to collaborate by introducing skills and tools for thinking systemically and holding conversations that bridge differences.
- Cultivate a viewpoint of responsibility for the problem (where it makes sense) and for how people are choosing to solve it.

CHAPTER 7

Facing Current Reality: Building Understanding Through Systems Mapping

The pioneering social psychologist Kurt Lewin said, "If you really want to understand something, try to change it." The link between insight and effective social change is critical. Too often people argue for solutions without a deep appreciation for not only how the current system operates, but also why. Too often the proposed solutions direct limited resources to solutions that make no difference in the long run or make matters worse.

The purpose of this chapter is to deepen your systems *understanding* by applying the storytelling devices introduced in chapter 4 to six cases. The cases focus on some examples that were introduced earlier—reorganizing responsibilities for K–12 education at the state level, criminal justice reform, and ending homelessness—and several new ones, such as improving rural housing and redesigning an early-childhood development and education system.

The emphasis in this chapter is on the first three tasks of facing current reality:

1. Identify people to interview about the history of the current situation and clarify what questions to ask.
2. Organize and begin to improve the quality of the information.
3. Develop a preliminary systems analysis of how different factors interact over time to support or undermine achievement of the vision.

Establish Systems Interviews

This first task will be more successful if you identify stakeholders with a wide range of views about the issue. This enables you to build a richer picture of how the system behaves and why. Most important, be sure to learn from stakeholders whose views you might otherwise dismiss. For instance, those spearheading a systems mapping project often exclude the most senior people involved in their issue because they believe they are too difficult to access or that they already know what those "at the top" think. Others too often excluded are administrative people (who might be affected by unwieldy processes and procedures), people whose work touches on the issue, and the ultimate beneficiaries of the work. In the cases we'll examine, these beneficiaries include homeless people, patients, and students. Learning from a diverse group of stakeholders not only builds understanding but also develops the relationships required to shift how the system operates.

For example, in the Calhoun County project to develop a ten-year plan to end homelessness, Michael Goodman and I interviewed fifty people representing a population of one hundred thousand. These included:

- Leaders responsible for the overall social and economic health of the county—formal and informal opinion leaders, representatives of state and local government, and leaders in the business and civic sectors.
- People who influenced policies affecting the likely causes of as well as solutions to homelessness.
- Service providers who interacted with and tried to help homeless people.
- Homeless people themselves.

Interviews, whether one-on-one or in small focus groups, are preferable to surveys because they enable you to uncover not only what people think but also the reasoning that leads them to their conclusions. Interviewing also builds more direct relationships. If necessary, a survey based on interview findings can then be used to gather information from a broader sample of stakeholders.

Over time, we have developed a list of interview questions designed to build systemic insights. These can be adapted for any issue, and used to jump-start the process:

- Consider first what has been happening around the problem you want to explore. Does it have a known pattern of behavior over time?
- If so, describe this pattern. Does it exhibit one or more of the following key variables over time:
  - Oscillations?
  - S-shaped growth?
  - Steeply rising, runaway growth?
  - Flat line, no growth?
  - Sharp ups and downs?
- State your definition of the problem by completing the following sentence: "Why has X been happening despite our best efforts to achieve a different goal?" "Why" rather than "how" questions ensure that your statements do not assume a solution. (For example, ask "Why does urban youth crime oscillate over time instead of continuously decline?" rather than "How can we toughen prison sentences to reduce crime even further?")
- What are the earliest antecedents of the problem? Also, describe any previous attempts to solve the problem. What was tried, by whom, and with what outcome?
- What is already working? How do people succeed or survive in the system as it functions now?
- How would this issue look from the viewpoint of senior decision makers? What factors or components would that level see? How do they think about the issue?
- How would other stakeholders, including ultimate beneficiaries, see the issue? What is important to each of them? How do they think about it?
- What other causes are affecting this system? What other effects (particularly those that are distant or unintended) does the system produce?
- What part of the issue is internal to my organization? What is a manageable chunk that relates to my position?
- In what ways do I or my organization create or contribute to the issue through what I/we say (or choose not to say), do (or choose not to do), or think?

- What is the apparent purpose of this system? In other words, what appear to be the outcomes of people's efforts? How is this different from what people really want?

It can also be helpful to tailor questions to the specific issue you want to address. For example, appendix B provides questions that were used to learn more about The After Prison Initiative, the Calhoun County Ten-Year Plan to End Homelessness, the effort to improve rural housing, and the Collaborating for Iowa's Kids project. The questions used for the Iowa project also represent a way of depicting an Accidental Adversaries relationship and were used in this case because the presenting issue was a difficult partnership.

Sometimes interviews are neither sufficient nor feasible to gain initial insight into an issue. It is still possible to draft a systems map using third-party reports drawn from such sources as written reports and meeting notes. For example, the initial systems map presented at the TAPI retreat was developed largely based on previous analyses of criminal justice reform supplemented by a few follow-up interviews. Preliminary insights into the root causes of civil war in Burundi came almost exclusively from written documents.

Organize Information

Here are four screens to use when organizing the information you gather:

- Listen for what is curious, confusing, or contrary among interviewees.
- Distinguish measurable data from how people interpret these data.
- Identify key variables, often thought of as critical success factors or key indicators.
- Look for recognizable story lines or archetypes.

The first screen involves noticing what concerns people or feels surprising to them. The focusing question itself, when phrased as "Why, despite our best efforts, have we been unable to . . . ?" evokes concern and curiosity. In their work to end homelessness, the leaders of Calhoun County wanted to understand how homeless people and vacant housing existed alongside each other. In seeking to improve rural housing, state and community leaders wanted to discern why some small towns appeared to have better housing stock than others. In the Collaborating for Iowa's Kids case, the two groups that shared a commitment to improve the state's K–12 education were curious about why it was difficult for them to work together.

The second involves learning not only what people think but also why. This means illuminating the assumptions and conclusions that people draw from what they actually do and observe. One common discrepancy emerges between what people think *should* happen as a result of their actions and what *actually* happens. For example, policy makers assume that drug busts reduce crime when in fact they reduce the supply of drugs, thereby increasing the likelihood that people will commit crimes to get the fewer and hence higher-priced drugs available. Different people also interpret the same data differently (the elephant analogy), and you can help develop a richer picture of the system by pointing out how each interpretation is valid under different conditions. In the homelessness case, local shelter providers saw the immediate benefits they provided homeless people, while those in other communities who understood best practices for ending homelessness understood the long-term costs of depending on a shelter system and the need to redirect resources to affordable permanent housing to solve the problem more fundamentally.

The third screen helps identify critical success factors or key indicators, and systems thinking extends this process in several ways. It looks for interdependencies among these factors instead of describing a laundry list that is difficult to prioritize. It acknowledges and integrates quantitative factors that are measurable with qualitative ones such as how people think and feel. It also translates often vague factors that are difficult to track over time. In the case of food aid, for instance, it can translate the vague factor of "strategy" into those factors that are traceable over time—such as the level of investment in local agriculture, which is a measurable form of strategy in ending starvation.

The fourth screen helps you identify recognizable stories, based on the systems archetypes or the Bathtub Analogy, in the information you gather. Discerning these stories can provide the basis for a comprehensive systems analysis. While they might not explain everything that is happening, they can often be expanded upon and modified to make identifiable sense of rich, complex, and sometimes conflicting information.

Don't expect, though, to find everything you need or want in the information you collect. It is important to recognize that people can learn a lot with incomplete information. All the data you want are rarely available, and yet people have to make decisions and take actions in any case. This process acknowledges that information gaps do exist and suggests two ways of responding to them. First, it always helps to approach systems change as a learner. People make hypotheses, take actions, and learn more from their actions so that they can make wiser choices the next time. Second, gaps in knowledge can trigger valuable research. Asking what is not known and how to find it are powerful questions that you might answer before or in parallel with taking new actions, depending on how much risk and time delay you are willing to accept.

Develop a Preliminary Systems Analysis

In this section you will see how people working on a range of issues used the tools of system archetypes and the Bathtub Analogy to deepen their understanding of why they were not succeeding despite their best efforts. These insights in turn created catalytic conversations about the nature of their challenges and what they could do to overcome them—topics that are the subject of the next three chapters.

You will learn how these tools served as effective building blocks for evoking more comprehensive and accurate stories. There are several ways to enrich primary plot lines with multiple plots and subplots that embrace more of the reality people experience in dealing with these issues. In particular, you will explore different approaches for balancing people's need for insights that are *complex* enough to embrace their many diverse perspectives with their need for those *simple* enough to be understood and acted upon.

This section highlights cases that exemplify at least one of the five major archetypes or the Bathtub Analogy introduced in chapter 4. You will learn about the contexts that brought people together and the insights they developed, both basic and enriched, that led to deeper understanding.

FIXES THAT BACKFIRE

Fixes That Backfire is the story of unintended consequences. It was one of the key themes in The After Prison Initiative introduced in chapter 3. You'll recall that this initiative brought together a hundred progressive leaders—activists, academics, researchers, policy analysts, and lawyers—for a three-day retreat to clarify what else could be done to help people suc-cessfully reenter their communities after incarceration and to redress the underlying economic, social, and political conditions and policies that contribute to making the United States the world's largest incarcerator among developed nations. It was convened by the Open Society Institute and organized by Joe Laur and Sara Schley of Seed Systems to help OSI gain more insight into the effectiveness of its funding in this area. Most of the participants at the retreat were Soros Justice Fellows or OSI grantees who competed for OSI funding at the same time that they shared a commitment to criminal justice reform.

Participants recognized that a primary driver of mass incarceration was not the level of violent crime itself, but the fear of others (especially people of color) that manifests as structural racism and the denial of these people's rights. While mass incarceration temporarily relieves the fear, as seen in balancing loop 1 (B1) in figure 7.1, it also results over time in more and more people who are eventually released from prison (95 percent of those incarcerated) with significant disadvantages caused by their prison experi-ence. The problems include: personal demons, poor health, disruptions to family and work life, restrictions to basic human rights (such as education, employment, housing, and voting), and failure of others to understand their challenges. These disadvantages in turn reduce access to such resources as education, housing, employment, and social support—thereby increasing the likelihood that many will commit another crime within three years and be sent back to prison. This can be seen in reinforcing loop 2 (R2) in figure 7.1. The US Bureau of Justice Statistics reports that "Among prison-ers released in 2005 in 23 states with available data on inmates returned to prison, 49.7% had either a parole or probation violation or an arrest for a new offense within 3 years that led to imprisonment, and 55.1% had a parole or probation violation or an arrest that led to imprisonment within 5 years."[1]

In addition, structural racism breeds more fear (R3), as evidenced in part by a more recent report by the Sentencing Project which found that "white Americans overestimate the proportion of crime committed by people of color, and associate people of color with criminality," and that for some crimes the overestimation was "by 20–30 percent."[2] The spiral of fear continues, even though the crime rate itself has actually been dropping for more than twenty years.

The participants at the retreat were engaged in many different efforts to reform the criminal justice system—not only those highlighted in figure 7.1, which emphasizes providing services to recently released prisoners. These efforts included developing innovative service programs, sentencing reform, prohibiting re-incarceration for technical parole violations, and fighting the growing prison lobby that favors mass incarceration for its own economic benefit.

So a second and richer diagram was developed to represent additional vicious cycles that need to be interrupted in order to shift the core dynamic

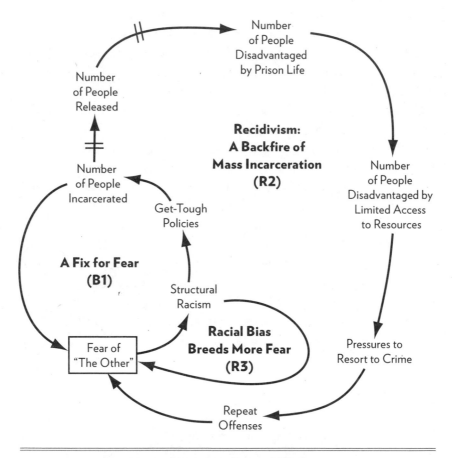

FIGURE 7.1 RACISM AND MASS INCARCERATION BREED FEAR. Although mass incarceration is a quick fix to people's fear of being victimized by crimes committed by people of color, it creates unintended consequences that further increase this fear over time. Modified from a diagram developed by Seed Systems for Open Society Institute

and to describe the more complex advocacy required to transform the system (see figure 7.2). Mass incarceration creates economic benefits that result in a prison lobby supporting this fix (R4 in figure 7.2). Tight restrictions on paroles to reduce the political risk of high caseloads result in more technical violations that send people back to prison (R5). A combination of fear and repeat offenses increases political resistance to innovations that could help formerly incarcerated people reenter society more effectively and prevent such offenses (R6).

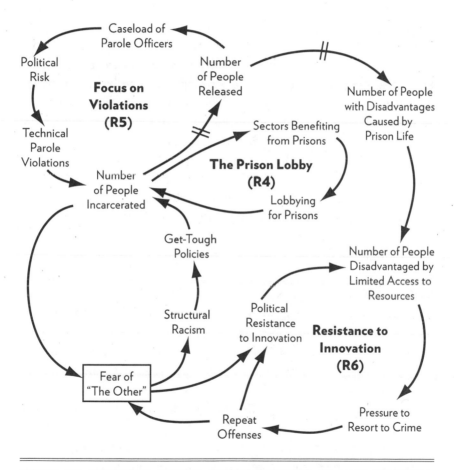

FIGURE 7.2 ADDITIONAL VICIOUS CYCLES PRODUCED BY MASS INCARCERATION. When the diagram in figure 7.1 is enriched to include more variables, it shows that three other vicious cycles—R4, R5, and R6—compound the Fixes That Backfire dynamic. Modified from a diagram developed by Seed Systems for Open Society Institute

These two diagrams (along with the final buildout "An Addiction to Prison" shown in appendix C, figure C.1) provided participants at the retreat with a comprehensive and shared understanding of their respective efforts in meeting common challenges. It enabled some to recognize and potentially develop more collaborative ways of reducing the fear that drives over-incarceration—a fear based in structural racism, not just the actual rate of crime. It gave others a way of communicating the societal costs of mass incarceration to policy makers in the hope of building a broader commitment to the underlying conditions and beliefs driving it.

SHIFTING THE BURDEN

Shifting the Burden is the story of unintended dependency. It is similar to Fixes That Backfire in that quick and seemingly obvious fixes to a problem tend to make the problem worse over time. The key distinction is that many of the people caught in a Shifting the Burden dynamic agree on what a more fundamental solution should be but have trouble implementing it.

This was the case in Calhoun County, where the community had made little progress in ending homelessness despite many years of effort. Those providing emergency and shelter services to the homeless had been meeting regularly for years but tended to work independently and compete for foundation and public dollars instead of collaborating.

The promise of state funding and local planning moneys motivated community leaders to take a different approach by agreeing to come together to develop a Ten-Year Plan to End Homelessness. They applied systems thinking to understand why homelessness persisted in the county and why they had been unable to implement best practices used by others around the country. These practices emphasized Housing First, where communities invest in developing safe, affordable permanent housing and encourage homeless people to live in it without first having to meet other conditions, such as substance-free living. This housing has easy access to support services for those who are mentally ill or suffer from substance abuse. For those forced to live on the streets for economic reasons, another part of the solution focused on education and economic development.

The initial systems map, depicted in figure 7.3, shows one of the critical realizations people made: The shelter and emergency response systems that both providers and the homeless depended on were quick fixes and in fact constituted part of the problem (B1). While these supports helped the homeless cope with their situation, they were also clearly temporary. People could stay in shelters for only a limited number of nights before they had to return to the streets. Some moved into the woods outside the main city of Battle Creek and survived there as best they could. Many got sick, especially in the winter, and ended up in the emergency room; some committed an offense that led them to spend a night in jail. Some spent nights sleeping on the couches of people they knew, but that solution also worked for only a limited time.

Many people close to the issue recognized at least some of the elements of the best practices used in other parts of the country—critical services, permanent housing, and employment (B2). However, they had been unable

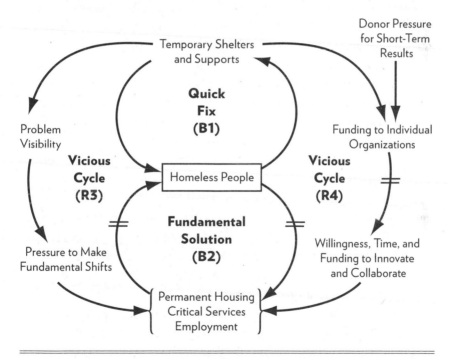

FIGURE 7.3 THE IRONY OF TEMPORARY SHELTERS. Because shelters and emergency supports temper the problem of homelessness, they also reduce people's motivation to implement a more fundamental solution that combines permanent housing, critical services, and employment. In addition, vicious cycles created by the quick fixes undermine the community's ability to implement the fundamental solutions even if it wants to. www .appliedsystemsthinking.com

to implement these practices locally and did not understand why. The community discovered that the emergency systems they depended on to help people cope with homelessness had actually led to negative unintended consequences that undermined people's ability to implement best practices required to end it.

The coping mechanisms described above decreased the visibility of the problem, thereby reducing the community's motivation to invest in best practices (R3). In addition, shelter and emergency service providers were funded for the work they did and had little incentive to experiment with new ways of doing things either individually or in collaboration with others (R4). For example, shelter providers were often funded on the basis of the percentage of beds slept in every night—even though full shelters are fundamentally antithetical to the espoused goal of ending homelessness. Donors often unwittingly contributed to the problem by renewing funding for those who succeeded in implementing quick fixes. The loss of problem visibility and reinforcement of existing funding streams made it difficult to mobilize investment in best practices, which in turn increased homelessness and the further dependence on temporary shelters.

This understanding led people to begin to make significant shifts in the community's shelter and emergency response systems. For example, in an unprecedented move toward collaboration, service providers collectively agreed to redirect their next round of HUD funding from one service provider's transitional housing program to a permanent supportive housing program run by another provider. Their insights formed the basis for the Ten-Year Plan that was accepted and funded by the state. As noted in chapter 5, in the plan's first six years of operation from 2007 through 2012, Calhoun County did a remarkable job of securing permanent housing for the homeless, especially in the face of the economic downturn of 2008–09. Homelessness decreased by 14 percent (from 1,658 to 1,419) *despite* a 34 percent increase in unemployment and 7 percent increase in evictions.

The TAPI case also includes a Shifting the Burden dynamic. Many participants were engaged in the fundamental work of resettlement, which they defined as strengthening civic institutions and infrastructure and facilitating cross-agency collaborations, as well as supporting former prisoners to lead productive lives. However, it was difficult to implement these solutions, and the reformers found it helpful to develop a shared picture of the obstacles they faced: the barriers to resettlement that undermined their work directly and threatened the safety of residents in the poor minority neighborhoods to which most men and women returned after prison, as well as the investment in prisons that could otherwise be directed toward the strengthening of institutions in these poorer communities. This picture is shown in appendix C, figure C.1.

Shifting the Burden is the story of *addiction*. The stories of both homelessness and imprisonment show how policy makers who want to protect society from addicts (homeless people suffering from substance abuse or drug addicts who commit crimes) can ironically become addicted to solutions that exacerbate these social problems in the long run. Shelters tend to divert money from affordable permanent housing, and get-tough policies take money away from the very kinds of programs and institution building that could more permanently reduce crime, the fear of crime, and most significantly the fear of "the other." The growing awareness of this consequence in the criminal justice system is prompting even some conservative public officials and their think tanks to question the value of mass incarceration and public policies such as the War on Drugs and mandatory minimums.[3]

LIMITS TO GROWTH

In 2011, a diverse group of state and local leaders gathered to explore ways of increasing the availability of affordable rural housing in South Dakota. The meeting was sponsored by the state's Rural Learning Center (RLC) and Dakota Resources, a nonprofit dedicated to economic, community, and leadership development. Participants included representatives of nonprofit economic development centers; government at the federal, state, and local levels; and developers, lenders, and Realtors from the private sector.

Housing is a critical success factor in the development of viable rural communities. Affordable and appropriate housing affects and is affected by other factors such as jobs, health, and the quality of education in these areas. The availability of this housing is especially important in drawing in young professionals to settle in and reenergize these communities.

Providing such housing was challenging since the costs to develop sufficiently attractive homes often exceeded what mortgage lenders in the current regulatory environment were willing to finance. Although the RLC had done a lot of work to write a detailed playbook for overcoming these obstacles, it was too much information for communities to absorb and take advantage of. Therefore, the purpose of the meeting was to gain insight into why it was so difficult to increase the viable rural housing stock and to target just a few high-leverage actions people could take to expand it.

I worked with the conveners to draft a systems map in advance of the meeting that might better illuminate the problem and possible solutions. Since certain communities had achieved more success in this area than others, it made sense to describe the dynamics in terms of Limits to Growth. We identified a core group of engines, or pumps, that could propel housing and economic growth but were weakened by balancing forces, or seepage, that reduced the effectiveness of these pumps. If the pumps could not produce housing and economic growth quickly enough, then the seepage would limit their effectiveness more severely. Therefore, it was important to prime the pumps powerfully and rapidly enough to overcome the negative effects of seepage.

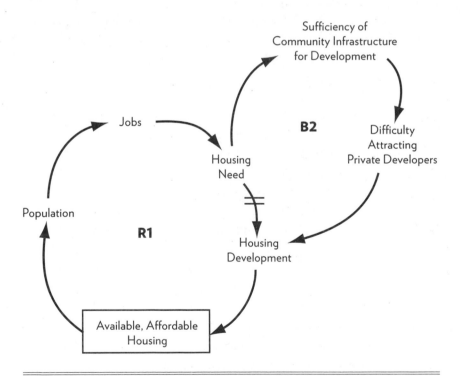

FIGURE 7.4 RURAL HOUSING: CORE LIMIT TO GROWTH. Without sufficient infrastructure for development, communities have difficulty attracting private developers to meet housing needs and thereby attract new people and jobs.

Figure 7.4 shows the core pump or engine of growth and the primary source of seepage or limit to this growth. Successful communities built affordable housing that attracted population and jobs, which in turn increased housing need and led to additional housing development (R1). They had done this in part by creating the local infrastructure for economic development. However, less successful communities did not have sufficient infrastructure to meet their housing needs, which made it more difficult for them to attract the private developers required to increase housing and grow (B2).

The communities identified other engines of growth as well, as depicted in figure 7.5. They recognized that housing development would in and of itself create jobs (R3 in figure 7.5), and that jobs would increase the tax revenues required to attract more jobs (R4). The availability of affordable

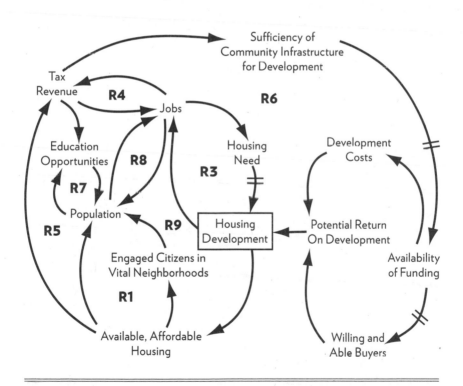

FIGURE 7.5 RURAL HOUSING: PRIMING ADDITIONAL PUMPS. Additional reinforcing relationships among jobs, housing, tax revenue, education, and population enhance community growth.

housing would also increase tax revenues directly (R5). Tax revenues could be used to strengthen the community infrastructure for development and increase available funding, which would in turn reduce development costs and increase the number of willing and able buyers, thereby increasing potential returns on development and housing itself (R6). Attracting

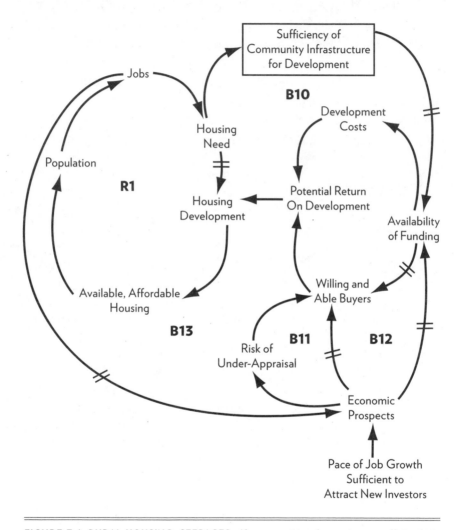

FIGURE 7.6 RURAL HOUSING: SEEPAGES. If communities do not invest sufficiently in their infrastructure for development, they will not bring in the funding required to increase potential returns on development. Moreover, if they do not create new jobs at a pace sufficient to attract additional investors, their perceived economic prospects will decline over time, thereby further limiting their ability to grow.

new people to the community would increase educational opportunities and attract more people (R7), and increasing jobs would have the same positive effect on population (R8). Finally, the availability of affordable housing would increase homeownership and engage citizens to contribute to building vital neighborhoods, thereby drawing in even more people to the community (R9).

Priming the pumps in figures 7.4 and 7.5 needed to occur quickly enough to overcome additional limits or seepages. Absent adequate infrastructure for development, communities could not attract the funding required to increase potential returns on development (B10 in figure 7.6). If they did not create new jobs at a pace sufficient to attract new investors, the community's perceived economic prospects would decline, thereby reducing the number of willing and able buyers over time (B11 in figure 7.6) and the availability of funding (B12). Moreover, weak economic prospects increased the risk of under-appraising the housing stock, thereby discouraging potential home-owners (B13).

In summary, the systems analysis enabled people to realize that committing to early investments in building or strengthening their communities' infrastructure for development was a critical first step to achieve greater growth.

SUCCESS TO THE SUCCESSFUL

Also in 2011 the William Caspar Graustein Memorial Fund (WCGMF) convened a multi-stakeholder process to create a blueprint for a statewide early-childhood development system in Connecticut that works for all children and families, regardless of race, income, or ability. The initiative, called Right from the Start, was designed and facilitated by the Interaction Institute for Social Change. It pulled together advocates, service providers, public agency staff, private funders, and community members/parents to explore and map what existed at that time, create a vision and underlying values for an equitable system, and identify key leverage points for change.

Several months into the process, I was asked to introduce the participants to systems thinking as a way of learning more about why their previous efforts to create a more equitable system had failed before jumping to propose new legislation and structures. Asking themselves why they had not succeeded until now despite their best efforts motivated many of them to ask more probing "why" questions as well. Since they were also committed members of the system, I encouraged them to consider how their own well-intentioned actions might contribute to the existing dynamics and what *they* might do differently in addition to recommending changes that the governor and other elected officials should make.

My encouragements to slow down and reflect more deeply before acting and to consider their own roles, however unintentional, in perpetuating the current system were confusing to design team members who viewed their charter solely in terms of rapidly producing recommendations about how others should change. A few chose to leave the project at this point while the majority took up the challenge to learn more before leaping ahead.

In a subsequent session I proposed a systems map to help them answer these questions. Since the core issue they presented had to do with inequity—the fact that children who struggled in school were primarily from disadvantaged black, Latino, and poor white families—I framed the underlying dynamic as Success to the Successful. Specifically, the education and income opportunities for people with higher incomes and political influence tended to increase over time as they successfully generated more resources for their children, while parents at risk (whether due to lower incomes, instability, and/or lack of service availability) tended over time to

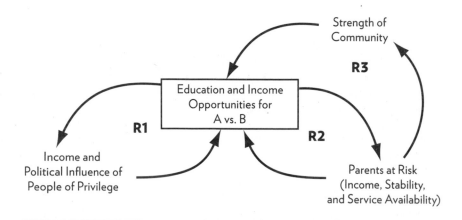

FIGURE 7.7 EARLY CHILDHOOD: THE RICH GET RICHER, AND THE POOR GET POORER. Children of parents of privilege are more able to experience the educational and subsequent income opportunities that lead to long-term success, while children of parents at risk are less able to generate these same opportunities over time.

find it increasingly difficult to provide their children with the early foundations they needed to become successful. Moreover, the strong community
bonds that poorer people often form to compensate for limited access to
other resources frayed further as they and their children struggled more
and more to succeed. These dynamics are summarized by loops R1, R2, and
R3 in figure 7.7.

Parents at risk had difficulty supporting their children to succeed for
several reasons, further fleshed out in figure 7.8. They had more problems
accessing affordable quality child care, which reduced their abilities to generate more income (R4 in figure 7.8). Because they themselves were more

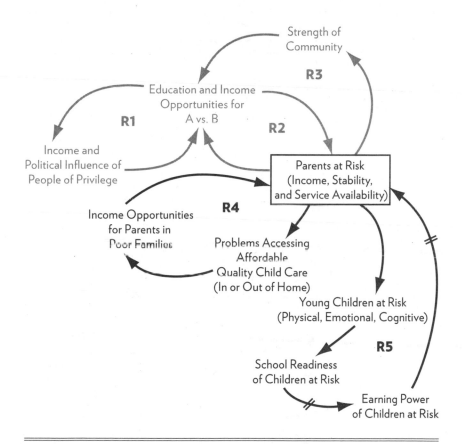

FIGURE 7.8 EARLY CHILDHOOD: THE GAP INCREASES. Parents at risk find it difficult to
generate sufficient income because of their parental responsibilities, and to provide the safe
and supportive environment their children need to thrive.

vulnerable, it was also harder for them to provide the physical, emotional, and cognitive supports required to help their children grow, especially at an early age when those supports are most necessary for healthy development. As a result, their children were less prepared for school and over time less able to succeed in school and thus generate the earning power required to escape becoming parents at risk themselves (R5).

Direct negative impacts on the children of parents at risk were exacerbated by the tendencies of parents of privilege to justify and hold on to their

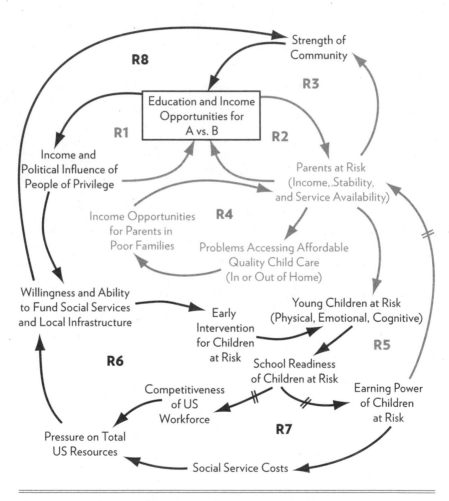

FIGURE 7.9 EARLY CHILDHOOD: THE GAP INCREASES EVEN FURTHER. The conservative nature of wealth combined with negative impacts of income inequality on public coffers undermine the prospects of children at risk even more.

wealth, thereby widening the gap between rich and poor children even further. In the United States, those in the top 20 percent income bracket give away only 1.3 percent of their income, while those in the bottom 20 percent give away 3.2 percent.[4] Dacher Keltner, a professor of psychology at the University of California–Berkeley and founder of the Greater Good Science Center, has found that, just as generous deeds tend to inspire greater generosity, the conservative effects of increased wealth seem to curb the instinct for helpfulness.[5] In addition, the weaker consumption and lower educational levels created by growing income inequality have reduced the availability of funds that can be invested in social services and local infrastructure. This next level of interrelated issues is mapped out in figure 7.9.

As collective willingness and ability to support these societal resources have declined, less money has become available for early interventions to help children at risk, which reduces their school readiness, and over time the competitiveness of the US workforce, thereby putting more pressure on total US resources and further reducing the ability to publicly fund social investments (R6 in figure 7.9). Reduced school readiness also leads to lower earning power over time, higher social costs, and an even greater drain on public resources (R7). In addition, when investments in social services and local infrastructure are cut, poor communities suffer even more, leading to greater inequality of opportunity (R8).

The participants also inquired more deeply into the underlying beliefs and assumptions that kept these dynamics in place. They realized that the problem of structural racism, whereby wealthy white people are viewed and treated more favorably than poor minorities, extended to assumptions about how to correct this imbalance. In particular, alternative solutions tended to favor the views of those in power that:

- Formal structures (such as new laws or institutions) should be emphasized over informal structures (such as social networks in poor communities).
- State control over the new system is more crucial than local control.
- Quantitative measures are more important than qualitative ones in assessing what works.

A profound insight for those who wanted to redesign the system was that they tended to subscribe to these assumptions themselves, often against their better judgment.

The participants then used their analysis to identify four leverage points, core commitments that continue to ground the work of everyone involved and inspire them to engage in the long-term work of systems change. These were:

- Address inequities and consciously empower the disadvantaged.
- Support local community action—recognizing how different service providers and families are already connected and how they should be connected.
- Pay attention to the whole child, which includes paying attention to the whole family.
- Begin intervening with children as early as possible.

The high level and supporting recommendations of Right from the Start have been incorporated into the guiding principles of the state's Office of Early Childhood, an agency that did not exist when the group began its work. They are reinforced by a communications campaign that includes videos with discussion starters for civic dialogue.

David Nee, the executive director of WCGMF at the time, observed that the work had "enabled our grantees to collaborate more effectively and target changes where we can have the greatest impact collectively. It helped all of us distinguish symptoms from the disease, appreciate the value of improving relationships among parts of the system more than improving subsystems, and understand that we are all part of the system and there is no one outside it." As a result of this understanding, the fund has strengthened its own will to deal directly with racism as an underlying problem.

ACCIDENTAL ADVERSARIES

The challenges experienced by the Iowa Department of Education and the state's Area Education Agencies (AEAs) are typical of organizations that want to be able to work together more effectively but find it difficult to do so. The Accidental Adversaries archetype provided a way for the two parties to surmount their differences in part because it gave them an opportunity to clarify how a stronger partnership would contribute to both of them. It also enabled them to see how the problems they experienced with each other were not intentionally caused by their counterpart. Finally, it showed them how they could capitalize on the benefits of partnership without unwittingly undermining each other.

Because Accidental Adversaries often fail to appreciate the value of their partnership at first, they benefit from learning to clearly articulate it. In this case, both sides acknowledged that the goals of the IDE were to provide guidance and governance for the overall education system across the state. They further agreed that the goals of the AEAs were to ensure quality educators in quality school systems for their respective geographic areas. Having affirmed each other's goals, they determined that a successful IDE contributed clarity of statewide direction and outcomes for students and educators to AEAs, and that successful AEAs would support the IDE by developing a coherent, equitable system across the state. The prospects of the IDE–AEA partnership are shown as a virtuous cycle (R1) in figure 7.10.

Having defined the value of their prospective partnership, they then uncovered how each party's actions to improve its own performance unintentionally undermined the effectiveness of the other party—outlined in

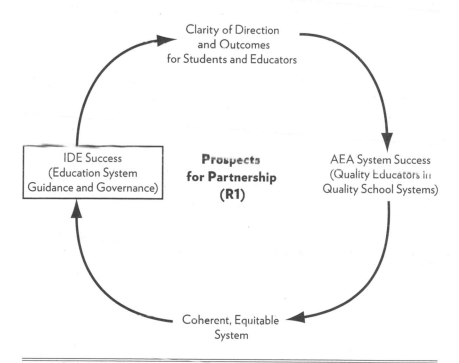

FIGURE 7.10 IDE-AEA (SYSTEM) RELATIONS: PROSPECTS FOR PARTNERSHIP. Both groups can benefit from the directional clarity provided by IDE and the coherence and equity ensured by AEAs.

figure 7.11. When IDE felt unable to provide adequate direction for the statewide education system, it designed and rolled out new programs to the AEAs (B2). While the existence of these new programs felt like progress to IDE, these initiatives made it more difficult for AEAs to allocate their limited resources with respect to the area initiatives they already had in place. In order to follow through on its own initiatives, each AEA would either customize IDE's programs to fit into the AEA's current efforts or disengage from the state rollout (B3). The independent AEAs' responses increased IDE's difficulties in working with inconsistent, low-quality, and disconnected solutions at the state level—reducing IDE success and prompting it to once again introduce new programs. The combined actions of the two parties resulted in the Accidental Adversaries vicious cycle R4.

The IDE–AEA story represents a common tension between headquarters and field units in any system. Headquarters tries to achieve its centralized goals by rolling out new initiatives to the field, and field organizations

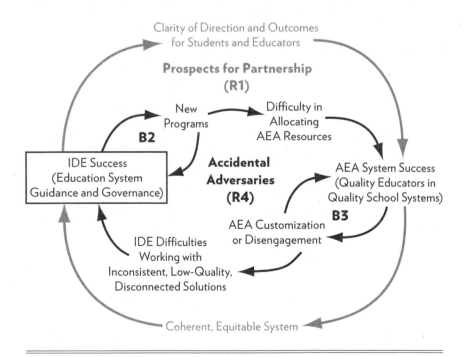

FIGURE 7.11 IDE-AEA SYSTEM RELATIONS: ACCIDENTAL ADVERSARIES. New programs created by IDE, and AEAs' efforts to customize or disengage from these programs, reduce both groups' effectiveness.

customize or disengage from these initiatives in order to better concentrate on existing efforts to serve their own respective clients. In this case the IDE and AEAs applied their new insights to create win–win solutions that supported each of their goals while ultimately better serving their mutual beneficiary—Iowa's kids.

THE BATHTUB ANALOGY

As introduced in chapter 4, the Bathtub Analogy is useful when attention must be given to the flows of factors as well as their levels. It is also helpful when an issue unfolds in stages over time. Michael Goodman and I used the analogy as a complement to causal loop diagramming in the ending homelessness case, and he has also used it recently to build on the rural housing case described in the Limits to Growth story mentioned earlier in this chapter.

In the homelessness case, we noticed that the problem of homelessness proceeded in stages. First, out of the overall population of Calhoun County, certain people were more at risk of becoming homeless than others— whether the risk grew out of poverty, unemployment, domestic violence, substance abuse, or some combination of these factors. While emergency support services were provided to people at risk, a certain number of people in this group became homeless. Homeless people also entered this community after being released from a nearby Veterans Administration hospital and prison. They moved into temporary housing and often recycled over time through different provisional locations: a shelter, under the bridge or in the woods, staying at others' homes, in the emergency room or local jail. Few of the people in temporary housing broke out of this loop and moved into permanent, safe, affordable, and supportive housing. We generated the Bathtub diagram shown in figure 7.12 to help them visualize the issue. The insight that the planning team got from that diagram is that the community needed to reduce the inflows into homelessness and accelerate the outflows from it in order to resolve the problem in a sustainable way.

One of the communities that participated in the South Dakota gathering to increase affordable rural housing was the community of Faulkton. Like many organizations and communities committed to managing a complex change, the Faulkton community group commissioned a study that proposed a lengthy list of recommendations to address its own needs for

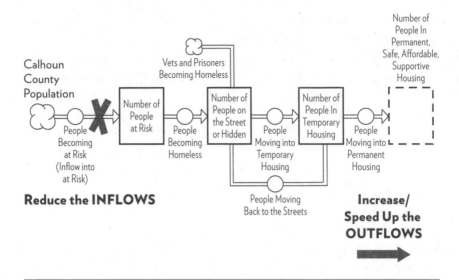

FIGURE 7.12 ENDING HOMELESSNESS IN CALHOUN COUNTY. In order to decrease homelessness, it is important to reduce the flow of people at risk and subsequently becoming homeless, and to increase the flow of people moving into permanent housing. www. appliedsystemsthinking.com

healthy and vibrant housing—seventeen in this case. Although the recommendations were categorized into multiple clusters, they seemed unwieldy given the community's limited resources.

Michael Goodman reviewed the report's findings (as part of a Dakota Resources Regional Systems Engagement Pilot funded by the Bush Foundation) and helped people focus on a few strategies that could have the greatest impact. He used the Bathtub Analogy to show the availability of four types of housing (see figure 7.13):

- New and sound housing that over time developed minor problems and became housing in need of minor repairs.
- Housing in need of minor repairs that, when left unattended, developed major problems and required major repairs.
- Housing in need of major repairs that, if left unattended, became dilapidated housing that needed to be rehabilitated.
- Dilapidated housing that needed to be removed to make way for new housing.

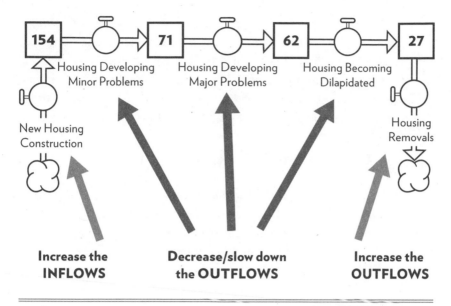

FIGURE 7.13 IMPROVING HOUSING QUALITY IN FAULKTON, SD. In order to improve local housing quality, it is important to speed up the flows of new housing stock and teardowns, while slowing down housing decay through repairs and rehabilitation. Innovation Associates Organizational Learning

The city realized that 28 percent of its housing stock was in need of major repairs or rehabilitation. They also determined through some simple simulation experiments using the housing stocks and flows that this number was likely to grow to 38 percent in twenty years if current trends continued. As a result, the community's seventeen recommendations evolved into just two primary strategies for meeting its housing needs:

- Decrease or slow down the outflows of good housing stock by repairing housing with minor and major problems, and by rehabilitating dilapidated housing where feasible.
- Increase inflows to the housing stock by building new housing and removing unrepairable dilapidated housing for new development.

Both Calhoun County and the community group of Faulkton benefited from using the Bathtub Analogy as a complement to causal loop diagramming and other forms of analysis. In the case of Calhoun County, the Shifting the Burden archetype pointed out that the community's unintentional dependence on temporary shelters as a way to help people cope with homelessness actually undermined its ability to end homelessness. In the case of Faulkton, the insight that emerged from the Limits to Growth archetype in the Rural Learning Center–Dakota Resources gathering about the need for sufficient community infrastructure for development evolved into a focus on two primary strategies that communities could implement to further develop their housing stock.

How to Balance Simplicity and Complexity

One of the challenges in developing a systems analysis is that it should be simple enough to be understood by people yet complex enough to capture the richness of their diverse perspectives and experiences. In this final section we will look at several ways of achieving this balance:

- One archetype—no additional loops.
- One archetype—additional loops all enriching the same story.
- Multiple archetypes—often using more than one diagram.
- Bathtub—either on its own or in combination with one or more archetypes.
- Interdependence mapping.
- Computer modeling and simulation.

Of the six cases described above, the Accidental Adversaries story that contributed to Collaborating for Iowa's Kids in figures 7.10 and 7.11 is the best example of the tremendous leverage that can be achieved using just one archetype with no additional loops. Even here it was helpful to develop the story in two steps (Prospects for Partnership and Accidental Adversaries) since the archetype embeds four loops: a virtuous cycle, two overlapping Fixes That Backfire, and a vicious cycle resulting from the overlap.

The affordable housing analysis for South Dakota and the early-childhood development and education system for Connecticut are examples of one archetype with multiple loops that all amplify and enrich the core theme. In the affordable housing case (figures 7.4 through 7.6), there are multiple growth engines trying to prime the pump (the R loops) that are undermined by multiple limits (the B loops)—yet the combined loops tell the same essential story of Limits to Growth. In the early-childhood case, there are multiple vicious cycles in figures 7.7 through 7.9 that aggravate the Success to the Successful dynamic.

The After Prison Initiative uses two archetypes to explain the reentry challenges facing formerly incarcerated people: Fixes That Backfire (figures 7.1 and 7.2) and Shifting the Burden (appendix C, figure C.1).

For those readers interested in resolving identity-based conflicts such as the Israeli–Palestinian issue, you can review three diagrams in appendix C (figures C.2, C.3, and C.4) that explain how such conflicts unfold from a combination of three dynamics: Shifting the Burden, Conflicting Goals, and Escalation.

The Bathtub Analogy represents another and sometimes more accessible way of understanding complexity. In the healthy and vibrant housing example presented to the community of Faulkton (figure 7.13), this analogy was extremely useful on its own. In the ending homelessness case in Calhoun County (figure 7.12), the Bathtub was used to complement the Shifting the Burden story described in figure 7.3.

For those who want a simpler form of storytelling, interdependence mapping offers another way of connecting multiple factors and the organizations associated with them. One simple variation asks people to first list the many factors affecting an issue and then draw relevant connections among the factors. Figure 7.14 represents such an example for ending homelessness.

Alternatively, people can clarify their interdependencies by drawing a linear input–output diagram or a Bathtub picture such as the one in figure

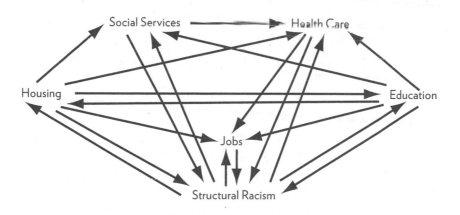

FIGURE 7.14 SIMPLE INTERDEPENDENCE MAP RELATED TO HOMELESSNESS. Interdependence maps are an easy way to raise the awareness of diverse stakeholders that their work is connected.

7.12. Once people draw this map, they can place their respective organizations on it by locating those factors or stages that their work most addresses. Stakeholders who see themselves on the map can then look at the bigger picture and reflect on the following questions:

- Whose work in this system do we support?
- Whose work in this system supports our own?
- What current relationships would it be helpful for us to improve?
- What new relationships would it be helpful for us to develop?

Since a system improves by strengthening the relationships among its parts, this simple exercise can be very effective in jump-starting the process.

On the other hand, computer modeling and simulation are helpful when the goals are to capture more complexity, validate that the systems map models known behavior, rapidly test different policy alternatives, paint different scenarios, and quantify the outcomes. These models incorporate both stock-and-flow dynamics as well as nonlinear and time-delayed feedback loops.[6] A user interface or simulator can be developed that enables stakeholders to easily explore and discuss the implications of various options.

By using some form of systems mapping to show the big picture, and engaging stakeholders to build or refine the map as much as possible, people learn to stimulate more productive conversations about the issue. The new conversations increase collaboration because they illuminate such critical factors as personal responsibility, interdependence, time delay and the difference between long-term and short-term impacts, unintended consequences, and local versus systemwide optimization. Designing and facilitating these conversations to build support for a systems analysis is the topic of the next chapter.

Closing the Loop

- Learning from a diverse group of stakeholders not only builds understanding but also develops the relationships required to shift how the system operates.
- Organize information by listening for what is curious, distinguishing measurable data from how people interpret these data, identifying key variables, and looking for recognizable story lines or archetypes.

- Stakeholders can use system archetypes and the Bathtub Analogy to begin gaining insight into why their chronic, complex problem exists.
- There are at least six ways to develop a systems analysis that is both simple enough to be understood by many people and complex enough to capture the richness of their diverse perspectives and experiences.

CHAPTER 8

Facing Current Reality: Building Support by Bringing the System to Life

Even though systems maps can build understanding, people do not necessarily embrace the insights they offer. For example, some participants in The After Prison Initiative retreat questioned the value of the map since those who in their minds really needed to change—in this case the policy makers who support mass incarceration—were not present. In Right from the Start and other situations, people ask how the insights can be communicated since they are presented in a visual language that is foreign to those not familiar with it.

The insights generated by systems maps can be difficult to accept and communicate. One reason is that the maps are a visual form of storytelling, yet not all people are visual thinkers. Removed from discussion about what they reveal, the maps can also appear abstract and impersonal, when in fact the stories they are intended to communicate are deeply human. Moreover, most maps challenge people to take more responsibility for the current system than they believe is necessary. Finally, even if people understand and accept the implications of the maps for themselves, they are often not sure how to convey this "foreign language" to stakeholders who have not been initially exposed to it.

You can meet these challenges by:

- Engaging people to develop their own analysis as much as possible.
- Surfacing the mental models that influence how people behave.

- Creating catalytic conversations that stimulate awareness, acceptance, and alternatives.

Engage People in Developing Their Own Analysis

Engaging people in developing their own analysis builds ownership for the work and increases its accuracy. If you are leading a mapping effort, it's important to proceed with that engagement goal in mind. When using causal loop diagrams, begin by drafting your own initial analysis based on interviews, focus groups, and previous written observations about the issue. (The maps in chapter 7 represent the types of analyses you might develop on your own or with a coach.) Then invite a small group of representative stakeholders, such as a steering or design committee, to comment on how the preliminary map both increases their understanding and can be strengthened.

Rather than present your findings from the outset, consider how this group might arrive at similar insights on their own. You could show them the template for a systems archetype that seems particularly relevant to their situation (such as Shifting the Burden, Accidental Adversaries, or one of the others introduced in chapter 4) and ask them to fill in the template to illuminate their story. Or you could write some of the key variables on Post-it notes and ask them to draw the most important cause-and-effect relationships between the variables. After having developed their own insights, people can more readily interpret, internalize, and improve your draft. If you are consulting to a project and cannot work with a small group first, then it helps to ask one or more people involved with the project to comment on the initial map and refine it before engaging a larger audience.

Presenting effectively to a larger group begins with anchoring the analysis with a story. For example, one story in Calhoun County revolved around a man who had cycled continuously through the community's various forms of temporary shelter for many years: official shelters, the street or woods, emergency rooms, jails, and the couches of those who knew him. People wanted to help this man get on his feet and find more permanent housing but seemed unable to do so despite their caring and efforts. Why was it difficult for the community to break the cycle of temporary shelters and ongoing vulnerability? What might they do differently to create a more permanent and satisfying solution?

Before offering a map that might help answer people's questions, it is important to position this diagram as a catalyst for conversation and learning—not as a definitive answer. Remind people that they will be asked to comment on any map presented to ensure it reflects their experience. Then, after the presentation, invite them to offer their feedback and suggestions, which I find easiest to record on flip charts in the moment and later incorporate into the map offline.

When preparing and presenting your map, practice reverse translation—converting the systems language used to develop the map back into English. In order to make the map accessible to people unfamiliar with systems language, it helps to:

- Refine the map first to remove all jargon and communicate the loops in everyday terms such as *quick fix*, *vicious cycle*, and *growth engine*. In the rural affordable housing case, reinforcing loops were described as "pumps" and balancing or limiting loops as "seepage" to evoke an agricultural metaphor.

- Introduce relevant archetypes as stories common to the human experience—not just to the people in this particular situation. This helps people both recognize the story and relieve any shame associated with being caught up in it.

- Honor people's best intentions by acknowledging what they *want* to accomplish through their actions. For example, they want to reduce starvation and believe that sending food is the humane and right way to accomplish this, or they want to grow their community and believe that building new housing is the key to attracting new people and jobs.

- Then show how these well-intentioned actions fall short because of factors and consequences, often longer-term, that have not been taken into account. As we learned earlier, for instance, those who prioritize food aid when trying to end starvation unintentionally undermine local agricultural development by driving down food prices. They also create a spike in population growth years later as young children saved by food aid become of child-bearing age themselves, often leading to greater famine due to the compromised local food system.

- Introduce the diagram in stages using "builds." For instance, in chapter 7 we saw several stories unfold in stages—including an Accidental Adversaries story that first showed the prospects of partnership, and then showed how the potential partners accidentally become adversaries. Each diagram has its own recognizable theme as the story expands and deepens.

A lingering question you might have at this point is "how do we know if the map is accurate?" One way to establish accuracy is to engage people in developing it using the guidelines in this section. Another is to ensure that the map answers the focusing question posed at the beginning of the process and explains why key variables have trended over time as they have. A third test is more visceral: A useful map often evokes palpable silence on the part of participants. The silence communicates many feelings in rapid succession:

- Humility that comes with recognizing that, in the words of Walt Kelly's Pogo, "We have met the enemy, and he is us."
- Despair from acknowledging that we can no longer keep doing what we have been doing and expect a different result.
- Hope that we can find a more effective way forward by thinking and acting differently.

Maps can evolve to not only enrich and expand people's understanding, but also sometimes to shift it profoundly. For example, a breakthrough insight for the TAPI participants was that the fear of being victimized by crime can drive behavior in the criminal justice system more than the level of crime itself. In a very different situation—the effort to rebuild civil society in Burundi after its 1990–94 civil war—NGOs that developed a systems analysis of the conflict determined that the driving factor in the war was not which tribe, Tutsis or Hutus, was in power, but the ability of an elite from either tribe to dominate the majority through ethnic manipulation.

Surface Mental Models

The comedienne Lily Tomlin once observed, "Reality is nothing more than a collective hunch." One purpose of systems thinking is to make our hunches explicit so that we can question and modify them to better achieve what we deeply care about. Indeed, my colleague Michael Goodman says, "Systems thinking is mental models made explicit." I think of it as a form of collective meditation: slowing down our thinking long enough to reflect on whether it serves us.

Our assumptions and beliefs drive many of the cause–effect relationships we experience. For example, consider if downsizing in an organization reduces or increases the morale of the remaining staff. The obvious answer is that cutting staff decreases morale because the other employees wonder, "When is the other shoe going to drop; will I be fired next?" or "My work-load has increased even further as a result, and I can't do it all." Alternatively, the morale of remaining employees might increase because they think, "We have finally removed the deadwood," and "I am considered important here and now have an opportunity to show even more of what I can do." Different assumptions determine if the relationship between downsizing and morale is undermining or constructive.

These perceptions are the next deeper level of the iceberg that explains current reality (see figure 3.2). You can clarify them by asking participants to look at the causal loop diagram and answer the question, "What are the key assumptions that keep this dynamic in place?" It helps to begin by identifying assumptions people hold about why current solutions or ways of doing things *should* work: "This quick fix should solve the problem because . . ." or "This engine of success should assure continued growth because . . ."

Encourage people to locate each mental model on a specific cause–effect link if possible and identify which stakeholders hold that particular belief. Placing mental models directly on the diagram brings systems maps to life.[1] For example, figure 8.1 shows the mental models various stakeholders had in Calhoun County that contributed to perpetuating homelessness. You can see that everyone held reasonable assumptions that unexpectedly contributed to the very problem they wanted to solve.

Sometimes people's assumptions are so deeply embedded in the overall dynamic that it is easier to list them separately. For example, participants committed to redesigning the early-childhood development and education

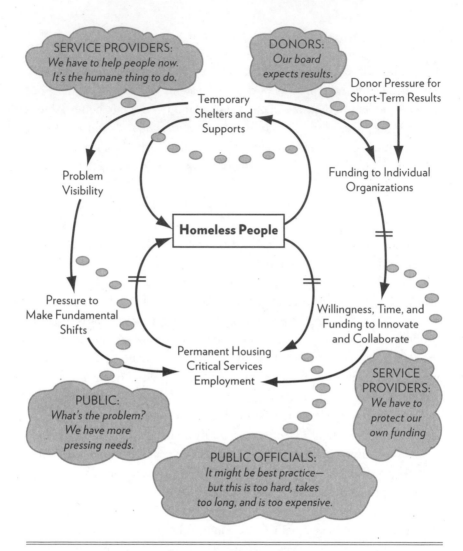

FIGURE 8.1 MENTAL MODELS IN CALHOUN COUNTY. Adding mental models to a systems map not only brings the map to life but also helps explain why certain dynamics persist. Bridgeway Partners and Innovation Associates Organizational Learning

system in Connecticut uncovered mental models that constrained how they thought about potential solutions to their problem—even though their own experience often contradicted the assumptions they held. They recognized that their beliefs reflected those of their political sponsors, who thought that:

- Formal structures (such as new laws or institutions) should be emphasized over informal structures (such as social networks in poor communities).
- State control over the new system is more crucial than local control.
- Quantitative measures are more important than qualitative ones in assessing what works.

At other times it can be illuminating to go beyond the space limitations of thought bubbles and describe how different stakeholders see the problem in greater detail. One dramatic example of this was provided by Jon Walz, a doctor and former fellow in the Kentucky Public Health Leadership Institute. Dr. Walz distinguished the respective views of patients, medical providers, and politicians who faced the challenge of how to end smoking. I find his work especially illuminating because he describes six different lenses or paradigms that summarize people's diverse views: defiance, fear, entitlement, desperation, ignorance or functional illiteracy, and recognition (see table 8.1 below). I suspect that these paradigms pervade people's views on other issues as well, although only one of them—the lens of recognition—promises a constructive solution.

Create Catalytic Conversations

The purpose of the systems map and inquiry into underlying mental models is to help stakeholders create catalytic conversations. Instead of re-creating familiar discussions about limited resources, who is to blame, and who else needs to change, these new conversations are designed to deepen awareness, cultivate acceptance, and develop new alternatives. People learn to see the system more comprehensively and usefully, accept their responsibility for the issue, and expand their views about what they might do differently.

DEEPENING AWARENESS

The systems map builds deeper awareness by helping stakeholders understand the often non-obvious interdependencies and longer-term consequences that undermine their best intentions. Adding mental models to the diagram or conversation surfaces unquestioned assumptions that

TABLE 8.1. MENTAL MODELS AROUND SMOKING CESSATION

	PATIENTS
Defiance	"I can smoke if I want to! My granddad smoked all his life and died from emphysema that he got from working in the mines."
Fear	"Nobody ever told me to stop . . . does smoking really cause cancer, Doc? I saw how my mom was when she passed and I don't want my kids to see me like that."
Entitlement	"I'm not going to stop until they buy my Chantix for as long as it takes . . . and if it makes me sick I'm gonna sue the fire outta them!"
Desperation	"I know it's killing me . . . but I just can't stop! Can't you hypnotize me or something . . . I'd give anything just to be able to breathe again."
Ignorance or functional illiteracy	"I went to that class but I couldn't understand half what they said."
Recognition	"I've wanted to do this for a long time and this time I'm going to do it for myself! I deserve it!"

Source: Courtesy of Jon H. Walz, DO, FAAFP

PROVIDERS	POLITICIANS
"I don't personally care if you stop or not! If you want to get well you'll throw those smokes away now!"	"We can't just turn over our entire budget to keep people from killing themselves! We've got to get control of this thing and put the responsibility where it belongs . . . on the tobacco companies!"
"Look, if you don't stop smoking, you are going to get worse and I'm afraid that I'll be criticized for your bad outcome after surgery."	"I think a man has a right to smoke if he wants to and I'm not going to let a little cancer get in the way of free choice! Why, I'll get run outta office if I vote to fund this cessation program!"
"If you don't stop, I'll be forced to refer you back to your old doctor!"	"The people in my community want CAT scanners and ambulances and that's what they're going to get. They don't care if a few kids smoke at school. I say, build the hospital now!"
"If you don't stop, there won't be anything I or anybody can do to help you when you really get sick!"	"If we fund this preventive health initiative, we won't have any money left to fund what we've already authorized and the state will go bankrupt!"
"There is nothing anybody can do for you! Cold turkey is the only thing that works and you just can't cut it! You're doomed!"	"None of this preventive stuff works anyway! Did you see the stats from California? They spent three and a half billion dollars on smoking education for kids and their lung cancer rates didn't change a bit! Are we going to waste our money like that?"
"I think you can do this if you really put your mind to it and let me point you toward some aids that are really proven to help! I'll work with you for as long as it takes!"	"This is a logical beginning to a long legacy of public health success here in Kentucky. By funding smoking cessation now, we will reduce the burden of disease now and in the future!"

people can now test. The following questions prompt catalytic conversations based on these inputs:

- What new insights have emerged about why the problem persists despite our best efforts to solve it? How do we see things differently?
- What is surprising?
- How is our group in part responsible, albeit unwittingly, for the issue?
- What challenges do these dynamics present?
- What new opportunities do they offer?

You can also develop more specific questions that promote inquiry into certain dynamics. For example, a school district that saw Success to the Successful as the root cause of its challenge to reduce the achievement gap between rich and poor students found enormous value in considering two questions:

- Why do some students who come from privileged families perform poorly in school?
- Why do some students from families with limited means perform well in school?

These questions prompted the district to realize there are certain critical success factors that enable *all* students to be successful—no matter what their backgrounds. These include: role models/mentors who believe in them, family and community support structures, resiliency and self-regulation, and honoring individual learning differences.

CULTIVATING ACCEPTANCE

With growing awareness come new challenges. Understanding why a system operates the way it does despite people's best efforts to improve it requires that they move from blame to responsibility, independence to interdependence, and short-term to long-term thinking. Accepting responsibility for your contribution to the current dynamics is the first step because it supports interdependence and long-term thinking. It is an act of self-empowerment—for, as Churchill observed, "The price of power is responsibility."

Two seemingly opposite stances support people in developing this acceptance: compassion and confrontation. You need compassion to recognize
that many problems are self-created, however unintentionally. Most people
do not want to increase starvation or perpetuate homelessness or destroy
the environment. It helps to assume that everyone is doing the best they can
with what they know at the time. On the other hand, it is equally important
to recognize that, at some level, people know not what they do and can
be their own worst enemies. They implement quick fixes and assume that
short-term improvements will last, or they build engines of success and
assume that this success will sustain itself indefinitely. They try to optimize
their part of the system under the assumption that this is the best way to
optimize the system as a whole.

Raising awareness and acceptance are not intended to replace blaming
others with blaming yourself. Rather they are designed to enable people
to achieve more of what they deeply care about—because it is ultimately
easier to change yourself than to change others (no matter how tempting
the latter notion). We know from experience that both compassion and
confrontation are vital elements of change—as when we talk about the need
for "tough love" or "ruthless compassion."

As a convener or facilitator of systems change, it is also important to foster confrontation without contempt. Confrontation means raising people's
awareness of their actions in service of helping them achieve what really
matters to *them as well as yourself*. Confrontation is founded on compassion. By contrast, contempt seeks to shame others into taking actions that
serve *you* independent of what is important to them. Contempt often shows
up for people in power, and it tends to evoke defensiveness rather than a
willingness to change.

One group that is frequently viewed with contempt is publicly elected
officials. Their actions often appear to be solely motivated by their desire
to get reelected, and they do that by appealing to the public's fears and
short-term interests rather than to their constituents' best interests. I raised
this concern once with a former congressman who had successfully served
his generally conservative district for twenty years, and then went on to
chair his state's nonprofit management association. I asked him how he had
navigated the tension between getting reelected and his own higher value
to serve the greater good. He told me that he viewed getting reelected as a
baseline, a necessary condition to serve his constituency, but not as an end in

itself. Similarly, as the famous management theorist Peter Drucker pointed out, profit is a necessary condition for business success but not its purpose, just as we need oxygen to breathe but breathing is not our purpose. I believe that many officials could be influenced on the basis of appealing to their higher values and thinking about how to help them get votes within that context, rather than by assuming that they are only motivated by immediate self-interest and belittling their views.

Lest this sound naive, think of leaders such as Martin Luther King Jr. and Nelson Mandela who confronted the power structure while at the same time empathizing with it. They appealed to the human dignity and higher aspirations of those in power and sought to partner with them on that basis. When collaboration failed as an initial strategy, they were willing to "fight fire with fire" but then return to collaboration when their adversaries demonstrated more willingness to act as partners.

DEVELOPING NEW ALTERNATIVES

People's willingness and ability to develop alternative solutions are founded on awareness and acceptance. When people become aware of why their own previous solutions have failed, and they accept their responsibility for these failures, then they are genuinely open to considering new ways of thinking and acting.

When you follow the previous steps in this chapter, people naturally begin to think about new solutions. It is fine to encourage fresh thinking at this point. At the same time, there is one other factor they need to consider before committing to alternative actions. The one other potential obstacle to moving forward is people's underlying intentions. Most of us have multiple commitments—for example, the desire to meet our more immediate self-interests and the desire to realize our higher aspirations. When these commitments compete, we will try to move forward only to be drawn in another direction and then feel pulled back and forth between the two.

The next chapter helps people examine these pulls more closely and resolve these conflicts. Supporting people to align their actions with their highest aspirations with considered awareness of how to also address their short-term interests is vital to releasing their full energies in service of lasting change.

Closing the Loop

- While systems mapping can increase people's understanding of why they have been unsuccessful despite their best efforts, it is not always easy to motivate people to act on these insights.
- To help people capitalize on the power of systems thinking, it is important to engage them as much as possible in developing the analysis.
- There are several ways to make systemic insights accessible to those who are not familiar with this way of thinking.
- Identifying the mental models that shape system dynamics not only breathes life into an analysis, but also provides a source of leverage for shifting these dynamics.
- The ultimate purpose of systems mapping is to catalyze new conversations.
- These conversations are most productive when they deepen awareness, cultivate acceptance, and develop new alternatives.

CHAPTER 9

Making an Explicit Choice

The broad-based teamwork involved in solving complex social problems requires, as we've now seen, aligning diverse stakeholders with a common public purpose—even though each may have different private agendas. In chapter 6, for instance, we saw the conflicts that almost always emerge in coalitions trying to end homelessness. Everyone has different primary concerns: Elected officials worry about containing costs to keep voter support; downtown businesspeople worry about keeping homeless people away from their storefronts; and shelter providers worry about filling beds to keep their funding. The approach recommended in chapter 6 to align these different interests is to establish common ground by clarifying people's shared aspiration and their initial picture of current reality. But the work doesn't stop there.

While developing common ground is vital, it can miss the even deeper challenge of aligning people with *themselves*. The diversity of concerns held by different stakeholders makes it difficult to not only align people with one another, but also to align each stakeholder's highest aspiration with their own immediate self-interests.

Most people are pulled between achieving what they most deeply care about and meeting short-term goals.[1] We want to realize our divine nature while also ensuring that we meet more basic needs such as economic security, belonging, and recognition. Even our desire to help others relieve their immediate suffering can conflict with helping these same people become independently secure and fulfilled over time. The subsequent question for those who want social change is how to support people to realize their highest aspirations, particularly when these diverge from their more immediate concerns. How do we help people make an explicit choice in favor of what they most profoundly want?

The answer is to connect people even more closely with both their aspirations and current reality by uncovering the bottom of the iceberg—the purpose that inspires them and, often by contrast, the purpose that shapes their everyday actions. By becoming more aware of both purposes, people can make a more conscious commitment to their highest aspirations with full awareness of the potential costs, not only the benefits, of realizing them. In order to align stakeholders most powerfully around their avowed purpose, it is important to help them make an informed choice to commit to this purpose in full light of what it might take to get there. Making this choice is pivotal to aligning people's energies in service of meaningful change.

You can learn to create this alignment by supporting people to take four steps:

- Understand that there are payoffs to the existing system—a case for the status quo.
- Compare the case for the status quo with the case for change.
- Create solutions that serve both their long-term and their short-term interests—or make a trade-off with the recognition that meaningful change often requires letting something go.
- Make an explicit choice in favor of their higher purpose by weakening the case for the status quo and strengthening the case for change.

Understand Payoffs to the Existing System

Systems are perfectly designed to achieve the results they are achieving right now.[2] At first glance, when we look at how dysfunctional existing systems can be, this premise seems absurd. For example, why would people design a system that perpetuates homelessness, increases starvation, or undermines children's abilities to learn? The answer that emerges from a systems analysis is that people are accomplishing something they want now, something other than what they say they want. They are receiving payoffs or benefits from the status quo, and they are avoiding costs of change.

Payoffs to an existing system include quick fixes that work in the short run to reduce problem symptoms and the immediate gratification that comes from implementing them. In systems that unwittingly perpetuate homelessness, some of the payoffs to the existing ways of working are reduced visibility of the problem due to temporary shelters that keep people off the streets or out of the public eye, reduced severity of the problem because some forms of shelter exist, good feelings on the part of both shelter providers and funders that they are helping people in need, and continued funding for the shelter system.

Costs of change that people prefer to avoid include financial investment, the discomfort of learning new skills and creating different work, having to act interdependently instead of independently, and being patient while waiting for investments to demonstrate returns over time. In the case of ending homelessness, some of the costs people avoid are investing in safe, permanent, affordable, and supportive housing; closing shelters or significantly shifting their mission and work; confronting fears on the part of citizens that they might have formerly homeless people as neighbors; and confronting fears on the part of homeless people that they might not be able to adjust to permanent housing.

The payoffs of the existing shelter system and the costs of changing it combine to yield a case for the status quo of helping people *cope with* homelessness. However, this case for the status quo actually undermines efforts to realize the avowed purpose of *ending* homelessness.

Compare the Case for Change with the Case for the Status Quo

The case for change includes the benefits of changing and the costs of not changing. These are often easier for people to clarify than the benefits of not changing and the costs of change. People have already been thinking about their vision for a desired future, and they can also imagine a negative future where the problems that concern them are not addressed.

In order to build the case for change, you can ask people what benefits would be derived from realizing their vision—benefits for their constituents and society as a whole, for their partners and other stakeholders, and for themselves. Those involved in trying to end (versus cope with) homelessness might respond:

- Reduced costs for the emergency response and social services associated with chronic homelessness—including shelters, hospital bills, and substance abuse treatment.

- Reduced unemployment costs for people who experience episodic homelessness because they have lost their jobs and ability to pay for housing.
- Ability to receive state and federal funds for meeting best-practice requirements to reduce homelessness.
- Positive feelings associated with providing people with permanent housing.

Then, to help people understand the costs of not changing, you can ask them to paint their nightmare scenario—to describe the worst that could happen if they do nothing differently now. For those same people working to end homelessness, the costs of not changing include:

- All the above costs continue to increase.
- Lost funding caused by failing to meet government requirements for implementing best practices.
- Lower quality of street life leading to economic decline.

In order to help people compare the case for change with the case for the status quo, it helps to complete a cost–benefit matrix, as shown in table 9.1.

The cost–benefit matrix helps people understand at an even deeper level why change is not occurring despite their best efforts. It depicts the often hidden case for the status quo—one that is currently strong enough to override the case for change and perpetuate the way things are.

Create Both/and Solutions—or Make a Trade-Off

People ideally want to have their cake and eat it, too: They would like to keep the benefits of the status quo while also realizing the benefits of change. Indeed, both/and solutions are preferable where they can be found, and there are a number of methods such as Polarity Management for creating those solutions.[3] In the effort to end homelessness, there are hundreds of community-based continuums of care throughout the nation providing housing and services for homeless people. Components may include: street outreach, emergency shelters (least permanent), transitional housing (supporting chronically homeless people to prepare to live in permanent housing), rapid rehousing (helping homeless people quickly move into permanent housing, usually in the private market), permanent supportive

TABLE 9.1. COST-BENEFITS MATRIX FOR ENDING HOMELESSNESS

Case for Change	Case for Status Quo
Benefits of change	**Benefits of not changing**
• Reduced costs of emergency response, shelters, health care, substance abuse treatment, unemployment	**(payoffs of status quo)**
	• Reduced visibility of the problem
	• Reduced severity of the problem
• Increased ability to receive government funding	• Good feelings of helping people in need
	• Continued funding for shelter system
• Positive feelings associated with providing people with permanent housing	
Costs of not changing	**Costs of change**
• Above costs continue to increase	• Investment in safe, permanent, affordable, and supportive housing
• Lost funding from failing to meet government requirements	• Closing shelters or shifting their mission and work
• Lower quality of street life leading to economic decline	• Confronting fears of ordinary citizens
	• Confronting fears on the part of homeless people

Source: Bridgeway Partners and Innovation Associates Organizational Learning

housing (permanent, affordable, safe housing combined with supportive services for chronically homeless people), and services only. There can be a place for all these alternatives as long as the overall system is incentivized to provide people with permanent housing as quickly as possible.

However, more often than not, people have to make trade-offs. They have to decide if what they aspire to is worth giving up at least some of what they have. As much as we prefer not to let go of anything to have even more, we also understand "no pain, no gain," "there is no such thing as a free lunch," and "investing now for the future." Not only do systems exhibit a tendency for better-before-worse behavior (for example, through quick fixes that undermine long-term effectiveness), but the reverse is also true. Things often have to get worse (or more difficult) before they get better. We have to let go of something such as comfort, security, and independence to have what we want even more. By contrast, the unwillingness to let go of such benefits to the status quo is the greatest obstacle to change.

Lyndia Downie, the president and executive director of Boston's Pine Street Inn, one of the most respected shelters in the country, realized that the inn needed to totally transform its mission in order to truly tackle homelessness.[4] She discovered that 5 percent of the homeless people in her shelters took up more than half of the beds on any given night, and that these were the chronically homeless who most needed permanent housing. Committed to Housing First, which centers on providing homeless people with permanent housing quickly and then providing services as needed, she convinced her board to transform the inn's mission from emergency shelter provider to real estate developer and landlord. She describes the "hard-to-stomach" decision for both the board and staff that involved closing some shelters and shifting those resources to buy homes instead.

The COO of a nonprofit committed to providing health care for the homeless in a major city participated with all the stakeholders working with the homeless in her area in an exercise to develop a systems map. After reviewing the map and her organization's place in the larger system, she challenged her president and board at a meeting later that day, "What might we have to give up as an organization in order for the whole to succeed?" I had never heard the question put so boldly before and realized that it makes an enormously powerful statement.

As in the case of the Pine Street Inn, sometimes the greatest challenge begins with letting go of one self-image and replacing it with another:

- The Area Education Agencies and local school districts in Iowa realized that they needed to give up their identities as being solely responsible for the students in their respective geographies. In order to improve education outcomes across the state, they needed to access the power of interdependence and let go of a measure of independence with respect to each other and the state Department of Education.

- The regulator of food safety in a major city learned that it was more effective when it shifted its role with restaurant owners from being an enforcer to being an information provider and educator.

- A county public health department increased its ability to improve the health of a poor community when it shifted its role from being an arm's-length expert to becoming the facilitator of a community-driven process.

Make an Explicit Choice

You can support people to let go more easily by first weakening the case for the status quo, and then strengthening the case for change.

A systems map naturally helps weaken the case for the status quo by showing how people's current thinking and actions tend to lead them away from achieving the purpose they aspire to. For example, the emergency response system to cope with homelessness unwittingly diverts attention and resources away from ending it. Separately optimizing parts of the K–12 education in Iowa undermines the state's ability to improve education outcomes for all its children. Depending on enforcement as a way to motivate restaurant owners to increase food safety makes it more difficult to achieve the cooperation required to do so.

Strengthening the case for change involves two steps that deepen people's connections with their highest aspirations. The first is more *receptive* in nature and supports people to stop and listen to what calls to them most authentically. Otto Scharmer describes this as *presencing* in his pioneering book *Theory U: Leading from the Future as It Emerges.*[5] He states:

> *Presencing—the blend of sensing and presence, means to connect with the Source of the highest future possibility and to bring it into the now. When moving into the state of presencing, perception begins to happen from a future possibility that depends on us to come into reality. In that state we step into our real being, who we really are, our authentic self.*

Presencing evokes a deep connection described by different names in various wisdom traditions. Scharmer describes it as an eco-centered view, one captured by the famous philosopher Martin Buber when he encouraged people to "Listen to the course of being in the world . . . and bring it to reality as it desires." Asking, "What is being called of us?" can lead people in a significantly different direction than one based on the question "What do we want to create?"—which risks focusing them on a more ego-centered place.

The second step in deepening people's connection to the case for change is more *active* in nature. It supports people to envision the ideal future that profoundly calls to them. The following guidelines for visioning are based on principles developed by Robert Fritz, a master of the creative process:

- Separate what you want from what you think is possible.
- Focus on what you want versus don't want.
- Focus on the results instead of the process.
- Include the consequences you want.
- See/experience the vision in the present.

I then ask people several questions to describe an ideal time in the future when the vision has been accomplished:

- How are the people you want to serve being served? What are they doing, seeing, feeling, hearing, and saying?
- How does serving these people contribute to other stakeholders and society as a whole?
- What is your group doing differently? What are you seeing, feeling, thinking, and hearing?
- What am I personally doing differently? How does realizing this vision serve my highest self?

Weakening the case for the status quo and connecting people more closely to the case for change through both deep listening and visioning help people make an explicit choice in favor of their highest aspirations.

What Can You Do When People Are Still Not Aligned?

While the four steps above stimulate alignment among diverse stakeholders, they do not guarantee it. One possible outcome is that you still cannot find common ground on which people want to build something together. In this case it helps to remember the alternatives proposed in chapter 6:

- Collaborate indirectly by legitimizing and addressing others' concerns, and then seeking to influence them through mutually respected third parties and/or to engage them at critical phases in the process.
- Work around the people you cannot work with.
- Work against them through such channels as advocacy, legislative policy, and nonviolent resistance.

It is also important to recognize that not everyone needs to agree at once on a new course of action in order for change to occur. Everett Rogers's famous study on the diffusion of innovations concluded that attitudes shift progressively through a population, and that the 15 percent who comprise innovators and early adopters can build sufficient momentum for others to follow.[6]

Another possible scenario is that people look clearly at the case for the status quo and the case for change—and deliberately decide to maintain the status quo with full appreciation of the future they are giving up on. This is certainly a valid choice, and I only encourage people in these cases to make peace with what they have—since they are now consciously choosing it. This means accepting all of current reality including its undesirable aspects since none of it is likely to change if they do not change themselves.

Closing the Loop

- It is difficult to establish common ground when people's everyday actions are not aligned with their highest aspirations.
- Helping people make an explicit choice in favor of what they most profoundly want is a pivotal stage in the change process.
- You can enable people to align their current behavior with their avowed purpose by supporting them to take four steps:
  1. Understand that there are payoffs to the existing system.
  2. Compare the case for the status quo with the case for change.
  3. Create both/and solutions—or make a trade-off.
  4. Make an explicit choice in favor of their higher purpose.
- You still have alternatives available when stakeholders do not align around a higher aspiration even after taking these steps.

CHAPTER 10

Bridging the Gap

The medical informatics unit of a large nonprofit health care system had a clear and compelling vision: assure that the most advanced knowledge about medical informatics be incorporated into the wider organization's clinical information systems. The unit's members included doctors and information system professionals with advanced degrees, and they all had a passion for the contributions they wanted to make. However, despite strong support from senior management, the unit faced multiple problems, including trying to convince a loosely knit confederation of hospitals to implement its ideas, making commitments to these hospitals and then failing to deliver high-quality systems on time, and burning out its staff in the process. People were overcommitted and underdelivering.

The challenge of work overload is not limited to people committed to social missions.[1] However, the very nature of having high aspirations, limited resources, and difficulty measuring progress leads nonprofit organizations to not only inspire people but also burn them out. It can feel like *anything* we do will move the ball forward, so we try to do *everything* we can with little heed to strategic focus and sustainable energy. The tendency to try to do too much is compounded when donors have unrealistically high expectations of grantees and grantees agree to these expectations in their efforts to compete for funds. Goals escalate, priorities proliferate and shift, quality suffers, and tensions mount as people fail to come through on agreed-upon tasks.

The good news is that systems shift not as a result of making many changes, but by sustaining focus on only a few changes over time. These changes are called leverage points because they leverage limited resources for maximum long-term impact. Instead of doing more and accomplishing less, organizations that target and implement such interventions actually achieve better results with less effort. Once people have clarified where they are and committed consciously to what they really want (with an appreciation of the trade-offs involved in getting there), they are ready to identify leverage points to bridge the gap and establish a process for continuing learning and outreach.

Identify High-Leverage Interventions

In her outstanding book *Thinking in Systems*, Donella Meadows identifies twelve leverage points in ascending order of impact.[2] This chapter offers a shorter and somewhat revised list that people committed to social change have found especially useful:

- Increase awareness of how the system currently functions.
- "Rewire" critical cause–effect relationships.
- Shift mental models.
- Reinforce the chosen purpose by aligning goals, metrics, incentives, authority structures, and funding to support it.

INCREASE AWARENESS

For those of us conditioned to take action, it feels strange to follow the advice ascribed to President Dwight Eisenhower, among others, who said, "Don't just do something; stand there." People who are impatient to define solutions and act upon them often fail to appreciate the power of a deeper insight into what is to catalyze change. As stated in the New Testament, "The truth will set you free" (John 8:32).

The framework used in this book emphasizes the stage of facing current reality—uncovering the non-obvious interdependencies that influence performance, appreciating the differences between the short- and long-term impacts of an action, recognizing your own responsibility for perpetuating the problem, and acknowledging the payoffs of the status quo—as a lever for change in and of itself. When the medical informatics group above realized that their difficulties in delivering quality products on time stemmed from their own tendency to overcommit, they acknowledged the importance of negotiating more realistic agreements with their internal customers.

The mapping tools in chapter 7 and catalytic conversation questions in chapter 8 offer numerous examples of people who began to shift as a result of becoming more aware of the system in which they were embedded. By way of review, some of the most powerful questions to raise awareness are:

- Why have we been unable to solve this problem despite our best efforts?
- How might we be partly responsible, albeit unwittingly, for the problem?
- What might be unintended consequences of our previous—and proposed—solutions?
- What are the payoffs to us of the current system?
- What might we have to give up for the whole to succeed?

We can also appreciate the power of awareness by its absence. As the celebrated historian Barbara Tuchman observes, the past is full of follies where leaders engaged in "the pursuit of policy contrary to the self-interest of the constituency or state involved."[3] Tuchman defined the criteria of folly as:

- "It must have been perceived as counter-productive in its own time, not merely in hindsight."
- "A feasible alternative course of action must have been available."
- "The policy in question should be that of a group, not an individual ruler, and should persist beyond 'any one political lifetime.'"

While her book focuses on four historical examples to demonstrate her position—the Trojan War, the rule of the Renaissance popes, the British loss of the American colonies, and the United States' disastrous Vietnam War—it is sadly easy to summon up recent examples as well. These include the Iraq War and our continuing hesitance to combat climate change. For example, in their recent book *The Collapse of Western Civilization*, historians Naomi Oreskes and Erik M. Conway uncover a pattern of denial of climate change that appears drawn from the tobacco wars: "Insist that the science is unsettled, attack the researchers whose findings they [the dissenters] disliked, demand media coverage for a 'balanced' view."[4]

REWIRE CAUSE–EFFECT RELATIONSHIPS

Rewiring means altering the cause–effect relationships that influence how people behave. In order to shift system dynamics and the patterns of behavior they produce in a sustainable way, some feedback relationships need to be created or reinforced to motivate new behavior or support what works. Others need to be weakened, broken, or even reversed to discourage reactive behavior and encourage more creative responses. In addition, time delays need to be shortened, lengthened, or tolerated if they cannot be changed.

One of the benefits of mapping reality in terms of archetypes is that we know a lot about rewiring common archetypal patterns. This section focuses on identifying generic leverage points for these patterns. It will emphasize the five common social change archetypes detailed previously: Fixes That Backfire, Shifting the Burden, Limits to Growth, Success to the Successful, and Accidental Adversaries—and then summarize the interventions for the other archetypes and the Bathtub Analogy. As in any organizational or community-building effort, it helps for the convener or facilitator to introduce the principles or even specific recommendations and then invite the input of a larger group of stakeholders before moving ahead.

People embedded in Fixes That Backfire have three options:

- Consider the negative long-term unintended consequences of alternative quick fixes, and choose a fix that appears to have none or at least fewer such consequences than the current one.
- Continue to use the quick fix if you must, but consider ways to mitigate its negative consequences.
- Uncover the root cause of the problem symptom that a fix is intended to address, and solve the underlying problem if possible.

For example, because mass incarceration leads to the eventual release of most prisoners with even more serious disadvantages and a high risk of recidivism, one alternative that is being addressed is modified sentencing. Alternatively, if imprisonment is necessary, it is important to actively design the prison experience as a place for reform (through counseling, education, job training, and continued connections with family) instead of punishment (with the attendant barriers to reentry faced by formerly incarcerated people). A third alternative is to invest in the development work that builds strong communities so that they do not become breeding grounds for despair and criminal behavior in the first place.

The three rewiring options for Shifting the Burden are:

- Reduce dependence on the quick fix.
- Increase investment in the fundamental solution by creating a vision of an alternative future that compels this investment over the long term.
- Where it is necessary to continue to use the quick fix while also working on the fundamental solution, design the fix in such a way that it builds toward this solution instead of undermines it.

One critical success factor in ending homelessness is reducing dependence on the temporary shelter system. In the course of transforming itself from a shelter to a Realtor, the historically successful Pine Street Inn now manages thirty-six 18-unit houses in Boston and neighboring Brookline and has more than half of its beds in homes instead of shelters.[5]. The approach is based on Housing First, a best-practice model of permanent supportive housing, designed to "Provide housing first, and then combine that housing with supportive treatment services in mental and physical health, substance abuse, education, and employment."[6] While shelters remain an important emergency measure even for the Pine Street Inn, efforts are under way there and in such nationally recognized communities as Columbus, Ohio, and elsewhere to position shelters as conduits to permanent housing instead of acceptable substitutes for it.

Conservatives often rightfully point out the tendency for public-sector organizations to become sources of dependency—a key indicator of Shifting the Burden dynamics. In the public health arena, several agencies have learned to recognize this dynamic and alter their policies as a result. For example, they increase food safety by partnering with restaurant owners to create food worker training systems instead of depending on their traditional enforcement role, and they create healthy communities by working with poorer communities to build local infrastructure over time instead of coming in as short-term experts.[7]

The keys to overcoming Limits to Growth are to:

- Anticipate potential limits even as you build engines of growth.
- Invest in overcoming these limits *before* they become a problem.
- If necessary, fund this investment from an existing growth engine even if it means growing more slowly.

Participants at the retreat on rural housing discovered that investing in infrastructure for community economic development needed to precede home building because this infrastructure was required to attract the funding and expertise of private developers. In order to sustain the effectiveness of its successful mosquito containment program, a public health department engaged health and other officials in neighboring communities to provide the personnel, equipment, and expertise needed to expand implementation beyond city limits.

Meeting the challenges of Success to the Successful is difficult because there appear to be few incentives for the more successful to give up their relative power. Some interventions that can work are to:

- Develop an overarching goal that links the achievements of A and B.
- Enable the more successful party A to recognize the negative impacts of inequity on it, including economic costs and social instability.
- Support party B to cultivate neglected sources of power such as tight-knit family and social traditions, numbers and hence votes, and moral rightness.
- Create systems that promote equity of opportunity and access.
- Invest in A and B based on their potential for success versus current performance.

For example, a community coalition in Eagle County, Colorado, was committed to a vision where all children, including those who came from very wealthy and poor families, were loved and successful. It discovered overarching educational goals by looking at two groups of outliers: kids who "have it all" (in the sense of coming from families of monetary privilege) but struggle in school, and other kids who succeed despite coming from families with limited monetary resources. Coalition members were asked to clarify what makes *every* student successful by considering two questions:

- What can be learned from both outlier groups about the resources required to succeed?
- What can be learned from these groups to increase all children's probabilities of success?

As a result of answering these questions, the coalition affirmed that several conditions were required for any and all children to succeed, including:

- Family and community support structures.
- At least one formative relationship with a teacher, advocate, or mentor in school.
- Challenge.
- Resilience and self-regulation.
- Honoring learning differences.

In the spirit of creating opportunities for all children in the county, the coalition named its work InteGreat! Bridging the learning gap evolved into an intention to meet these success conditions for every student.

In the case of helping children in Connecticut get support well before they entered school, there were several arguments put forth about the need to involve successful businesspeople in promoting early-childhood development and education based on their interests in developing an educated workforce and expanded consumer base. However, the long-term nature of these benefits made it difficult to engage them. On the other hand, participants did recognize the importance of cultivating neglected resources in poor communities, including social networks, their ability to influence local politics, the moral legitimacy of equal opportunity, and the power of communicating qualitative—not just quantitative—descriptions of what works.

In the United States, progressive taxation of both income and inheritance, as well as antitrust laws, are among the levers used historically to level the playing field and create equal opportunity. Education, albeit a longer-term solution, is also a key lever, along with the early-childhood development work energized by such initiatives as Right from the Start.

There are three high-leverage interventions that transform Accidental Adversaries into productive partnerships:

- Clarify or remind both groups how they can benefit from partnering with each other.
- Point out that the ways in which they have undermined each other are unintentional; each group has simply been trying to succeed on its own without considering the impact of its solutions on the other.
- Support both groups to look for win–win solutions: those that increase each group's success while also supporting—or at least not undermining—the other group's performance.

The education organizations in Iowa were able to collaborate more effectively with each other when they applied these three steps to improve relationships between the state Department of Education and the Area Education Agency system, individual AEAs, local school districts and individual AEAs, and school districts and the state Department of Education.

Table 10.1 summarizes how to rewire the cause–effect relationships that produce other archetypal behaviors:

Finally, when using the Bathtub Analogy, it is important to think about changing flows in order to change stocks or levels. For example, if a community wants to reduce the level of homelessness, it needs to decrease the inflow of people becoming homeless and increase the outflow of people moving into and remaining in permanent housing. Alternatively, to increase the level of affordable housing, it needs to increase the inflow of new construction and renovation and to decrease outflow of housing deterioration and ultimate neglect.

SHIFT MENTAL MODELS

People's mental models govern many of the critical cause–effect relationships that shape system performance. When considering an alternative to the existing quick fix in Fixes That Backfire, it is important to challenge the mental models that determine the use of this fix in the first place. For example, members of the medical informatics unit needed to question their belief that the best way to build support for their work was to raise expectations about the exciting benefits their software would deliver.

In order to reduce dependence on the quick fix and build support for investing in the more fundamental solution in Shifting the Burden, it helps to determine the assumptions that favor use of the quick fix and those that discourage implementation of the fundamental solution. In Calhoun County this meant challenging the assumption that shelters were an important part of the solution to ending homelessness, and that Housing First was not cost-effective.

Potential partners seeking to reverse the Accidental Adversaries dynamic need to learn that the other party's disruptive actions are not intentional, as the agencies in Iowa discovered when they mapped out the structure they were caught in.

TABLE 10.1. REWIRING OTHER ARCHETYPAL DYNAMICS

Dynamic	Interventions
Vicious Cycle (simple)	• Identify a weak link—one governed by people's assumptions rather than hardwired • Redirect the causal factor in this link by creating a new goal • Clarify corrective actions required to achieve the goal • Implement reinforcing actions to sustain momentum
Ineffective Balancing Loops	• Make continuous improvements based on a long-term vision • Exercise patience in the face of time delay, or find ways to reduce it without creating negative consequences • Base strategic solutions on clear and agreed-upon goals and a shared picture of current reality (including why it exists)
Drifting Goals	• Hold the vision • Reduce performance shortfalls by sustaining corrective action instead of lowering the goal
Competing Goals	• Look for a higher goal that encompasses the competing ones • If achievement of both goals is mutually exclusive, commit to one • If not, determine different corrective actions that lead to the accomplishment of both goals
Escalation	• Create full awareness of the structure and its costs • Look for ways for all parties to achieve their objectives • Invite the parties to agree to a balanced situation • Slow the rate of escalation • Deescalate unilaterally
Tragedy of the Commons	• Bring attention to the collective costs of individual actions • Focus on the greater common good or vision • Manage the common resource, perhaps through an agreed-upon higher authority • Close off the commons to allow it to replenish or regenerate
Growth and Underinvestment	• Anticipate and invest in advance of limits to be ready to respond to demand • Recognize critical performance standards and hold to them

Source: Adapted from Innovation Associates Organizational Learning

Surfacing, questioning, and testing people's beliefs and assumptions are essential skills in rewiring these relationships. Because many people tend to identify closely with what they think (as Descartes observed, "I think therefore I am"), helping them shift their mental models requires careful planning. The following five-step process can help:

1. Surface and respect current beliefs.
2. Ask, "Do these mental models help us achieve what we want now?"
3. Stimulate alternative views.
4. Develop a vision of what we want now and the mental models that would support it.
5. Conduct and learn from experiments.

Chapter 8 described a highly effective way to surface mental models and simultaneously bring systems maps to life by adding thought bubbles to key cause–effect links. Communicating respect for people's beliefs is important because beliefs are based on their past experiences—however incomplete or potentially outdated they are now. As Robert Fritz observed, asking people if their beliefs are *true* or not is a weak question because the answer is always yes. The medical informatics unit had received a substantial budget to develop new products based on the benefits they promised, and for many years shelters had provided the most humane places where homeless people could sleep.

Fritz recommends substituting *utility* for *validity* as the basis for assessing existing mental models. In other words, it is more productive to ask people if their current beliefs are *useful* or not. Do their beliefs enable them to achieve more of what they want now? For example, members of the medical informatics unit realized that, while making grand commitments to the system's hospitals had raised enthusiastic support for its work, overcommitting had also led to high stress, lower quality, and reduced credibility over time. Continuing to base agreements on potential benefits was no longer useful and in fact had become counterproductive.

You can also help people reduce their attachment to current beliefs by introducing alternative views. Getting the whole system in the room and creating catalytic conversations across diverse stakeholders inevitably surfaces and prompts different ways of thinking about the issue and the opportunities. The pioneering thinker on organizational learning Chris Argyris recommended that people seek disconfirming views—actively recruiting evidence that demonstrates how their mental models are incomplete or no longer accurate. For example, the fields of appreciative inquiry and positive deviance challenge assumptions that obstacles cannot be overcome by redirecting attention to how people have already succeeded in the face of similar challenges.[8]

You can also reduce attachment by proposing alternative interpretations of known facts—for example, by questioning if a demonstrated short-term improvement is sustainable or a shortfall is attributable to a lack of patience rather than a failure to demonstrate measurable progress. Systems maps stimulate new thinking by surfacing unintended consequences of people's actions, differentiating the short- and long-term outcomes of their solutions, and uncovering often non-obvious interdependencies among different parts of the system.

If current beliefs do not support desired results, then the next step is to develop a vision of what people want now and a set of beliefs that support movement in this direction. New supportive beliefs are not necessarily the opposite of existing ones. In fact, they are often more nuanced. For example, most members of the medical informatics unit realized that they wanted credibility—not just enthusiastic interest—for their work. They agreed to test a new assumption that success depended on making realistic agreements with potential beneficiaries instead of simply raising expectations. In order for leaders in ending homelessness such as Lyndia Downie to be successful, they have to believe that most (not necessarily all) homeless people want to live in a permanent home. The agencies that were Collaborating for Iowa's Kids needed to believe that partnering with each other to improve educational performance would produce better results than acting independently.

New beliefs will only be sustained to the extent that people can validate them based on new actions. This is why experiments and prototypes are so important. They can not only reinforce a new idea by showing it works in practice, but also modify and improve upon the idea by learning what does not work. Members of the medical informatics unit experimented with new ways of negotiating realistic agreements with hospital staff, and many were surprised to discover that, by challenging unrealistic expectations and developing alternative proposals, they were received as more professional rather than less.[9] Downie and the staff of the Pine Street Inn proved that most homeless people want a more secure home by providing permanent, supportive housing and then determining that 96 percent of chronically homeless stayed in their new home after one year and reported that "the camaraderie is terrific."[10] The Iowa Department of Education and Area Education Agencies worked together on improving early literacy—a high-leverage initiative in K–12 education—to demonstrate the benefits of partnering.

REINFORCE THE PURPOSE

Once people make a conscious commitment to their espoused purpose and identify leverage points, they often need to reassess the extent to which current goals, metrics, incentives, authority structures, and funding streams support or undermine the achievement of this purpose. For example, the steering committee in Calhoun County based the goals in its strategic plan to end homelessness on the leverage points identified in the systems mapping process. Three of these goals focused on the developmental processes of engaging the community, creating collaboration among providers and stakeholders, and establishing informed and aligned funding approaches. The other three addressed the fundamental solutions of providing access to quality, safe, permanent affordable housing; services to ensure stable housing; and permanent employment and education opportunities. Leading funders around the country have shifted the conventional metric for shelters of "increasing bed utilization" to "reducing length of shelter stay prior to a move to permanent housing."

The nationally recognized Community Shelter Board (CSB) of Columbus, Ohio, and surrounding Franklin County also demonstrates how these principles are applied.[11] CSB is a collective impact organization with twenty partner agencies founded by a powerful business community to end homelessness. One of its key early distinctions and levers was that it positioned itself to coordinate and directly oversee most of the community-wide funding dedicated to this outcome. CSB has used its financial clout to define metrics and incentives that are designed to optimize the whole system instead of its constituent parts. For example:

- Shelter providers are measured by the number of people housed and how quickly—not by bed utilization.

- Shelter and direct housing providers for families must work closely together to earn their money because shelters are incentivized to keep the number of days that families spend in shelters low, and direct housing providers must get referrals from family shelters to meet their goals of housing 70 percent of families.

- CSB runs a unified system for leasing up its permanent supportive housing units that assures that priority is given to those people with high vulnerability and who are also high utilizers of the mental health system. Instead of having waiting lists for housing, it has an integrated system with the mental health board, the housing authority, and CSB to assure that the neediest people are housed.

CSB is also an example of a backbone organization, which John Kania and Mark Kramer identify as one of the critical success factors for communities to achieve collective impact.[12] One of the benefits of backbone organizations is that they provide the authority structures necessary to focus the work of ongoing decision making and implementation. The participatory processes described in this book so far encourage shared responsibility and consensual decision making. Grounding the purpose in actionable tasks then requires that people authorize specific organizations or groups to decide and act on behalf of the whole.

Decision and accountability charting are tools that can be used to focus decision making and project implementation, respectively.[13] Both function under the premise that people are more likely to support processes that limit their ongoing involvement if they have been involved consensually in the previous work and are now asked to agree on how to set the limits going forward.

Decision charting assigns a few people to make specific types of decisions on behalf of the collective. The exact number of decision makers is as small as possible to make decisions efficiently and as many as necessary to ensure quality and support for implementation. Roles in addition to the decision makers include those with approval or veto power (identified only if required to meet legal, political, or fiscal obligations), those needed to support implementation of the decision, consultants to the decision, those who have to be informed of the decision, and a manager who ensures that the decision is made in a timely way.

Accountability charting uses similar role definitions with one exception. Instead of decision makers and a manager, there is one group or individual vested by the collective to drive implementation of a specific goal or project on their behalf. All other roles are the same as those used for decision charting.

Establish a Process for Continuous Learning and Outreach

In order to be effective in the long term, organizations need to follow through on their implementation efforts with a process of continuous learning and outreach. While identifying leverage points often entails an initial reallocation of existing resources (such as where people place their intention, focus, and time), this ongoing journey involves learning from experience, expanding the resource pool, and scaling up what works. Let's explore each of these.

Even beyond changing paradigms or shifting mental models, the ultimate leverage point identified by systems thinker Donella Meadows is transcending paradigms, which she defines as the ability "to keep oneself unattached in the arena of paradigms, to stay flexible, to realize that no paradigm is 'true,' that every one, including the one that sweetly shapes your own worldview, is a tremendously limited understanding of an immense and amazing universe that is far beyond human comprehension."[14] I think of this as the importance of establishing a process of continuous learning. Our visions evolve, current reality changes (hopefully in the direction we want), and new information and conditions emerge. The best we can do is to clarify what we want, plan how to proceed, take action, and learn from what happens.

Continuous learning at the local level involves:

- Extensive and ongoing inclusion of stakeholders.
- A clear strategic plan with specific projects in support of the plan.
- A strong focus on data to support evaluation against the goals and metrics.
- Quarterly and annual evaluations that inform updates to the plan.

These are the four strategies used by CSB, which considers its commitment to continuous learning to be the foundation of its ongoing success.

As for expanding the resource pool, CSB taps additional resources through its extensive partnerships and fund-raising ability. In addition to being constituted as a partnership of twenty different local agencies, CSB is actively involved in national efforts to end homelessness. It is a member of the Leadership Council and participates in many national conferences and webinars. It also seeks out other communities for technical assistance, such as by learning shelter diversion strategies from New York City. CSB's roots in the business community motivate it to keep learning from the private sector about such areas as how to engage government, develop business strategy, and improve the efficiency of its new single adult system (by incorporating supply chain techniques using a business's Six Sigma Black Belts). The organization also remains closely connected to the public sector.

CSB's strong fund-raising ability is based on its continual emphasis on inclusion, hard data, and measurable performance improvement. In addition, HUD has recently granted the organization unified agency status, which enables it to bring more programs under the CSB umbrella and help those programs focus on results and outcomes in order to continue funding.

Scaling up what works can be thought of as strategies to engage many people and institutions beyond the relatively few key people you might have involved so far.[15] For example, in her Soul of the Next Economy Initiative, social architect Pamela Wilhelms emphasizes the importance of targeting three different groups: consumers, voters, and investors.[16] In its collaborative research with several nonprofits dedicated to expanding the impact of the social sector, the national association of funders Grantmakers for Effective Organizations (GEO) cites four broad strategies nonprofits can adopt to scale up their work:

- Expanding the reach of a successful program in the same or different locations.
- Spreading an idea within a certain area or system—geographic, organizational, or professional.
- Increasing the number of people or places that use or apply a new technology, practice, or approach.
- Ensuring that ideas become embedded in policies and hence new behaviors pursued by a government body, corporation, or other institution.[17]

GEO then translates these into four practices that grant makers can follow to support all of these approaches:

- Provide flexible funding over the long term.
- Fund data and performance management capabilities.
- Support capacity building and leadership development.
- Support movements.

GEO emphasizes the importance of scaling up with flexibility, ensuring that there is room to adapt what has been learned in one place to new contexts. I think of this as expanding the process of learning what works instead of seeking to impose specific solutions. Moreover, as in the four-stage process of applied systems thinking, an effective learning and outreach process needs to address both *internal* changes—those that shape people's personal intentions, thinking, and actions—as well as *external* changes in the beliefs, policies, and regulations that govern their collective behavior.[18]

In their research on reframing the art of helping, Pulitzer Prize–winning authors Nicholas Kristof and Sheryl WuDunn see extensive opportunities for scale-up in adapting business competencies to what is traditionally defined as social or public-sector work.[19] Integrating the strengths of business to increase society's abilities to help its vulnerable populations can come in multiple forms, such as:

- Funding the development in nonprofits of such business infrastructure and skills as improved marketing, information systems, and personnel management (as also suggested by GEO).
- Social impact bonds that accrue private-sector investments to fund public-sector innovations.
- Social enterprises that use a profit-making model to achieve a social as well as economic bottom line.
- Impact investing that funds new social enterprises from charitable donations.
- Large company investments in developing and underdeveloped nations that build new markets to meet the needs of the poor as well as employee morale.

Kristof and WuDunn also point out the power of engaging faith-based organizations and creating social clubs whose purpose is to serve others in need. They encourage secular and religious forces to collaborate more, especially in the area of advocacy, to accomplish far more than each can separately to fight against the common enemies of humanity. They also cite the power of secular giving communities or circles that cultivate the desire of people in groups to make a meaningful difference.

How to Integrate Multiple Interventions

With many interventions to choose from, the question remains: "Where do we start?" The leverage points proposed in this chapter are presented in a logical order, whereby:

1. Awareness uncovers interdependencies to be rewired.
2. Rewiring is supported by understanding and then shifting the mental models that influence key cause–effect relationships.
3. Reinforcing the purpose facilitates implementation of changes in connections and assumptions.
4. Continuous learning and outreach enable people to make necessary course corrections in systems that are ultimately too complex and dynamic to control.

It can also be helpful to create small, more implementable changes early on as long as they are positioned within a long-term strategy (a method used by Houdini in designing escapes from his seemingly inescapable traps!). Finally, one way to integrate multiple interventions into a dynamic strategy is to organize them into causal loops that feed forward instead of backward. Designing such systemic theories of change is the focus of the next chapter.

Closing the Loop

- Systems pivot around leverage points, a relatively few key coordinated strategies sustained over time that produce significant long-term improvement.
- Effective interventions shift the pattern of behavior produced by the system's dynamics in a sustainable way.
- Four high-leverage interventions are to:
  - Increase awareness of how the system currently functions.
  - Rewire critical cause–effect relationships.
  - Shift mental models.

- Reinforce the chosen purpose by aligning goals, metrics, incentives, authority structures, and funding to support it.
- Organizations need to reinforce the implementation of high-leverage interventions with a process of continuous learning and outreach.
- This ongoing journey involves learning from experience, expanding the resource pool, and scaling up what works.
- You can integrate multiple interventions into a clear strategy by following the natural progression introduced in this chapter and then designing a more specific systemic theory of change.

PART THREE

SHAPING
THE FUTURE

CHAPTER 11

Systems Thinking for Strategic Planning

The emphasis in this book so far has been on applying systems thinking *retrospectively*—to determine why people have not been able to achieve the results they want despite their best efforts. Insights into deeply embedded dynamics and leverage points reduce the likelihood that they will repeat past mistakes or "reinvent a broken wheel." It is also easier to test the validity of a systems map based on known history because the analysis should be able to predict what has happened until now.

However, being able to identify leverage points does not necessarily help people organize their thinking about how to move forward. For example, once the agencies committed to Collaborating for Iowa's Kids understood why partnering with each other had been difficult, they still needed to clarify a way forward that would enable them to take advantage of their insights. It is important to connect leverage points into a coherent path forward—one that links and sequences interventions over time, accounts for time delay and long-term as well as short-term consequences of possible actions, and shapes intentional reinforcing and balancing feed-forward relationships into the future.

Moreover, there are times when people want to create something new and do not necessarily have the benefit of hindsight. While it might be easy to conceive of what needs to be done going forward, it is not so easy to integrate these critical success factors into a coherent strategy. For example, when an oversight group committed to increasing the health of its rural county developed its first strategy road map, it realized that the many strategies, tactics, interim outcomes, and end results that people identified did not form a clear path of action. Questions of scope, priority, timing, time delay, unintended consequences, and sustainability of proposed improvements still remained.

Sometimes organizations are simply overwhelmed by too many choices and too few resources. The problem is compounded by the high expectations people place on themselves and a tendency to assume that doing more automatically leads to accomplishing more. Organizations may have more programs than they can support and no clear way of rationalizing among them. They may have what feels like a laundry list of priorities and no way of deciding among them without getting caught in dysfunctional conflicts about organizational turf and who gets what piece of the pie. By contrast, when the management team of a major child welfare agency sought to make sense of the organization's multiple programs and support functions, it recognized the need for a coherent strategy that would justify—and if necessary prune—its many commitments.

In all of these cases, it can be helpful to apply systems thinking *prospectively*—to create a road map going forward that accounts for the complexity of having to navigate so many interdependent factors over time. The circular road maps produced by planning systemically have several benefits over more common linear input–output models. They:

- Incorporate the dynamics of reinforcing and balancing loops into cause–effect relationships that feed forward—thereby representing how social systems actually behave and unfold.
- Clarify a path for optimizing the relationships among the parts of the system instead of the parts themselves.
- Integrate multiple success factors into a logical and sequenced set of actions over time.
- Take time delay into account.
- Include plans to make both short-term and longer-term sustainable improvements.

By incorporating these features, circular theories of change create pictures that quickly communicate a lot of readily understandable information. The maps have been very effective in fund-raising, orienting and engaging other stakeholders quickly, and providing clear navigation for many people over long periods of time. For example, the officers of the new Food & Fitness program at the W.K. Kellogg Foundation created such a map to sell their initial nationwide proposal to the board. The map summarized the program's intentions and strategic plan so clearly and succinctly that the board took the unprecedented action of approving their proposal at its very first review meeting. A group from northeast Iowa that subsequently developed a comparable map for their regional Food & Fitness project (see figure 11.9, later in this chapter) still uses the map to orient their work more than five years after creating it.

So in moving to prospective systems thinking, it is time to introduce two core theories of systemic change and describe how people used them to develop strategic plans that meet these three challenges:

- Organizing the leverage points determined by a root cause analysis of chronic, complex problems.
- Integrating the numerous critical success factors required to create something new.
- Streamlining choices among too many programs or priorities.

We'll also look at some guidelines for refining your theory over time to accommodate new information and changing conditions.

Two Systemic Theories of Change

My colleague Michael Goodman, a co-developer of the system archetypes, noted that there are two distinct branches of archetypes—those based on reinforcing feedback and those based on balancing feedback.[1] The former describe stories of success compromised by limits, and the latter tell stories of improvement disabled by negative consequences. There are two corresponding core theories of change using the systems thinking tools in this book.[2] The first seeks to amplify success, while the second seeks to correct a shortcoming and achieve a goal.

The Success Amplification story begins with one or more reinforcing loops—the factors of success that build on one another over time to create more success (R1 in figure 11.1). You can learn what is already working in the system through informal anecdotes or by using more formal approaches such as appreciative inquiry or positive deviance.[3] Alternatively, people may first list what they perceive to be critical success factors or, as in the case of articulating the prospects for partnership between the Iowa Department of Education and the state's Area Education Agencies, describe how a core success loop should work (for an example, refer back to figure 7.10). In order to ensure sustainable growth, it is also important to plan beyond the initial engine of success. The success theory needs to also consider possible limits to initial improvements (B2 in figure 11.1) and anticipate how these limits might be overcome by generating new engines of success over time (R3). Finally, it is vital to identify key time delays in cultivating and sustaining success.

If people cannot easily identify successes to build on, or if the prospects for creating a positively reinforcing dynamic are tenuous, it makes more sense to develop a systemic theory of change that bridges the gap between

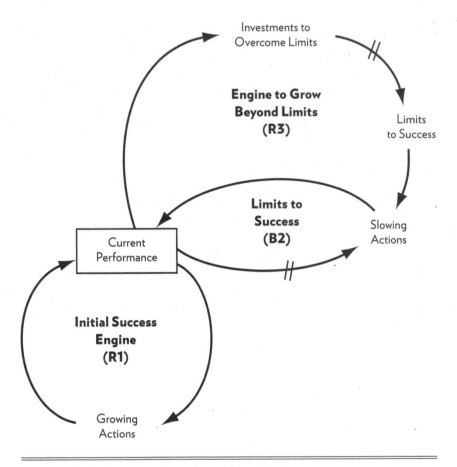

FIGURE 11.1 THE SUCCESS AMPLIFICATION THEORY. This theory clarifies how to build on existing successes, anticipate limits to even greater success, and create new engines of growth over time.

reality and a desired goal or vision. The Goal Achievement theory, mapped out in figure 11.2, begins with one or more balancing loops that identify the correction(s) required to close this discrepancy.[4] In order to determine what corrections are likely to be effective, it helps to first clarify the underlying structure that causes the gap (B1 in figure 11.2). Kathleen Zurcher points out that such an exercise is asset-based instead of deficit-based as long as the discrepancy is driven by a strong vision. In addition, because of the time delays usually required to close the gap, it is important to exercise persistence and stay the course in making the corrections (B2). At the same time, if progress is not occurring even after taking time delays into account, it helps to gain perspective on the reasons for the shortfall and rethink the nature of the challenge (B3).

Because people have a tendency to take the pressure off as they make progress toward achieving a goal, it is equally important in developing a systemic theory of Goal Achievement to define how people will sustain and reinforce improvements over time. Sustainability is a major source

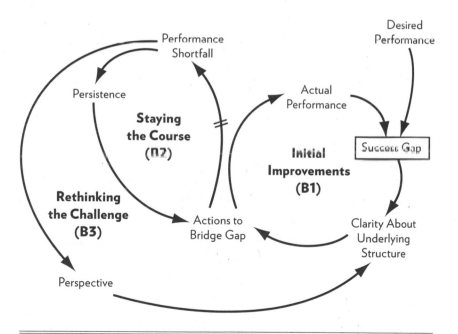

FIGURE 11.2 GOAL ACHIEVEMENT THEORY: MAKING INITIAL IMPROVEMENTS. This theory identifies initial improvements to accomplish a desired goal and recognizes the importance of both staying on course and rethinking the challenge to be effective.

of concern for funders who want to ensure that beneficiaries are able to continue if not expand their work without ongoing financial aid. The need for reinforcement is further evidenced by the emphasis in the quality movement on continuous improvement and observations by high performers that continuing to work hard is essential to sustained success.

Figure 11.3 provides a template for three engines of reinforcement. R4 encourages people to expand their aspirations as their actual performance increases. By experiencing their initial dreams come true and affirming new possibilities, they can raise their goals. By raising their goals, they can become clearer about new growth opportunities and take actions to realize them (R5). For example, growing actions available to the social sector include fund-raising based on demonstrated successes, investments in organizational capacity, partnerships that enable each organization to do what it does best while contributing to a larger result,

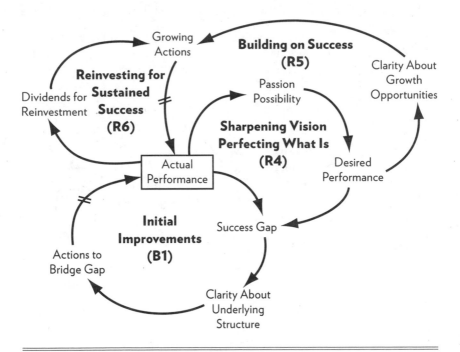

FIGURE 11.3 GOAL ACHIEVEMENT THEORY: REINFORCING IMPROVEMENTS. The second part of the Goal Achievement theory emphasizes the need for continuous improvement through sharpening the vision, cultivating additional growing actions, and reinvesting success dividends.

creative ways to scale up successes, and advocacy for policy change.[5] Finally, improvements in actual performance can redirect money previously required to treat problem symptoms toward new investments in the growing actions above (R6). For example, the annual savings of ninety-five hundred dollars per person achieved by keeping people in long-term housing in Massachusetts (thereby reducing the use of emergency rooms and temporary shelters) can be reinvested in programs that prevent or end homelessness.[6]

The next three sections give examples of how these core systemic theories of change provided road maps for organizations and communities committed to improving K–12 education, public health, and child welfare.

Organizing Leverage Points

The following two cases demonstrate how planning groups developed strategic road maps based on systems analyses of their core problems. The first revisits the agencies working to increase Collaboration for Iowa's Kids and uses a Success Amplification approach. The second introduces a coalition in a rural county committed to increasing public health, especially for its most vulnerable populations, and describes a Goal Achievement strategy.

AMPLIFYING STRENGTHS IN THE COLLABORATION FOR IOWA'S KIDS

Developers of the Collaboration for Iowa's Kids based their theory of success on the core reinforcing loop that described how the Iowa Department of Education and Area Education Agencies would capitalize on their prospects for partnership. The theory they developed and used as their road map is reproduced in figure 11.4. The central loop was the core theory underlying their partnership (as previously shown in figure 7.10). The right reinforcing loop provided a framework for AEAs across the state to increase coherence and equity within their own system, and the left reinforcing loop detailed how to increase effective collaboration between the IDE and AEA system. Although the theory does not explicitly map potential limits to success and ways to overcome them, developers did address these factors in their joint planning. They identified major implementation

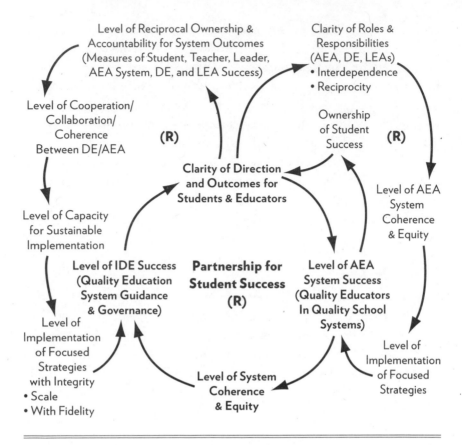

FIGURE 11.4 AMPLIFYING SUCCESS IN COLLABORATING FOR IOWA'S KIDS. This map was used by the Iowa Department of Education and the state's Area Education Agencies to explicate their theory for building and sustaining an effective partnership. The central loop describes broadly how collaborating benefits IDE and the AEAs, the right loop focuses on increasing coherence within the AEA system, and the left loop emphasizes improved implementation by both parties. Collaborating for Iowa's Kids: A Partnership Between the Iowa Department of Education and the Iowa Area Education Agencies

challenges to redesigning the AEA system and five strategies for overcoming these obstacles.

ACHIEVING THE GOAL OF A HEALTHY COMMUNITY

The Eagle County Public Health and Environment Department (ECPHE) shared a related commitment to the InteGreat! coalition introduced in chapter 10. It wanted to increase health equity in the county by focusing on the health of the county's most vulnerable populations, including children living in poverty, and incorporating the social determinants of health into all community decision making. ECPHE also engaged a group of diverse stakeholders, including some of those participating in InteGreat!, and developed a Goal Achievement theory of change. Their theory came out of a desire to discern the root causes of poor health in the county using a systems analysis and then design a strategy that would effectively integrate the identified leverage points.

The group described a common dynamic that I would call Treading Water, where strong forces dragging people down are barely compensated for by equally strenuous efforts to help them stay afloat. It identified five vicious cycles that led to the deterioration of the financial stability and health of vulnerable populations over time. The group also listed more than twice as many initiatives (in the form of balancing loops) that the community was taking to help break these cycles. The most powerful insight was that, despite the fact that there were many organizations providing services to improve the health of vulnerable populations (directly and indirectly), their efforts were significantly hampered by a failure to work together. Like oars in a crew shell that do not dip in the water simultaneously, this lack of coordination wasted resources and made it even more difficult for both providers and their intended beneficiaries to move forward and break the vicious cycles of poverty and sickness in the community.

This insight surfaced the need for several high-leverage interventions:

- Developing a map of current community health assets to both communicate the availability of these assets to vulnerable populations and identify opportunity gaps to be bridged.
- Increasing collaboration among service providers.
- Engaging vulnerable populations as equals and developing an effective support system for the community's undocumented immigrants, who make up a significant portion of the people at risk of getting sick.
- More effectively engaging the people who influence these individuals—including their families, employers, educators, churches, and political leaders.

- Better planning for the built environment, such as quality affordable housing and accessible recreational opportunities.

The advisory group also affirmed the need to continue to work toward
longer-term poverty reduction strategies such as increasing the number of
living wage jobs and the availability of affordable and effective health insur-
ance. Finally, they recognized the importance of shifting mental models in
the community that reinforced the status quo. These included:

- Classism, racism—"*They* are the kind of people who . . . aren't
  motivated to take care of themselves, learn, et cetera."
- "We can't change because of limited resources."

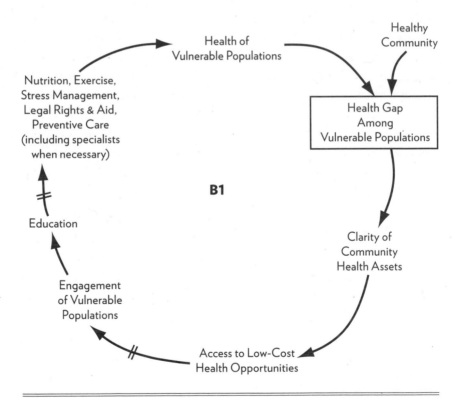

**FIGURE 11.5 CORE IMPROVEMENT TO INCREASE HEALTH OF VULNERABLE POPULA-
TIONS.** The core theory of improving community health focuses on developing a map of
the community's health assets as a way of engaging and educating the population, especially
those most at risk, about the many existing resources they can use to improve their health.

- "One size fits all."
- "The system is too big to change."
- "Health care can't be fixed."

The group first integrated the leverage points into several core balancing loops. These are shown in figures 11.5 to 11.7. B1 in figure 11.5 emphasizes the critical role that community health asset mapping has in improving access to low-cost health opportunities, which in turn increases the effective engagement and education of vulnerable populations, thereby increasing the level of their healthy activities and overall health as a result.

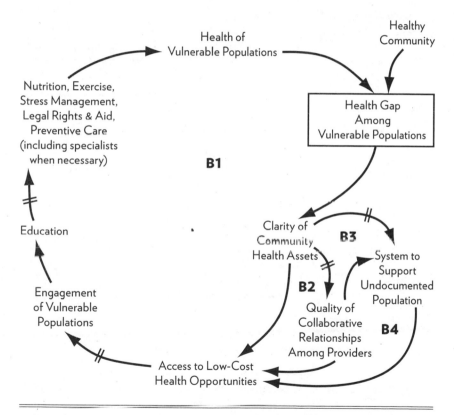

FIGURE 11.6 INCREASING COLLABORATION AMONG PROVIDERS. The group recognized the pivotal impact of increasing collaboration among service providers to both increase access to low-cost health opportunities and develop a support system for the county's undocumented population.

In figure 11.6, B2, B3, and B4 show how community asset mapping contributes to increasing collaboration among service providers and enables them to develop a support system for the county's undocumented population.

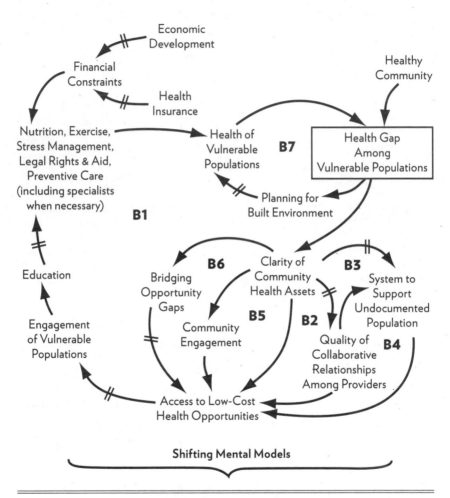

FIGURE 11.7 ADDITIONAL STRATEGIES TO INCREASE HEALTH OF VULNERABLE POP-ULATIONS. Several additional strategies were identified to build on the core ones of asset mapping and service provider collaboration: increasing community engagement, bridging opportunity gaps surfaced by the community health asset map, and taking health issues into account in planning for the built environment.

The advisory group also identified several additional strategies as shown in figure 11.7. B5 focuses on the importance of community engagement. B6 captures the value of asset mapping to identifying and bridging opportunity gaps. B7 affirms the importance of planning for the built environment.

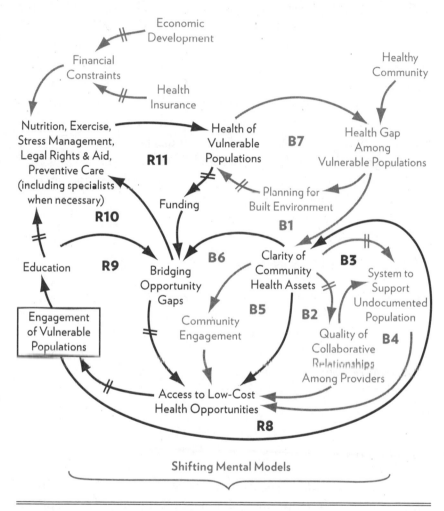

FIGURE 11.8 REINFORCING PROGRESS TOWARD A HEALTHY COMMUNITY. The group identified four ways to reinforce the initial improvements in community health over time: developing even greater clarity about the community's health assets by engaging vulnerable populations, illuminating and bridging opportunity gaps through the educational process, taking advantage of new opportunities to foster healthy activities, and using improved health indicators to attract additional funding.

Factors in the upper left corner of the diagram that impact the target population's financial constraints are also shown. Finally, in its line at the bottom of the diagram, the group acknowledged the importance of shifting mental models throughout the change process.

The group then identified four ways to reinforce the improvements described above, as shown in figure 11.8. R8 notes the value of engaging vulnerable populations to develop further clarity about the community's health assets. R9 shows the importance of designing an education process that supports people, ideally in small groups, to identify additional opportunities to improve their health. R10 emphasizes that people who identify new opportunities are more likely to take ownership for implementing their own self-care initiatives. R11 explains the power of using demonstrable successes in improved health to generate additional funding.

The two cases above describe how people have developed systemic road maps for change based on root cause analyses. The next section provides examples of developing these road maps based on assumptions about critical success factors without such analyses.

Integrating Success Factors

The first case in this section uses a Success Amplification strategy to build on strong existing relationships in a region to improve people's food and fitness. The second case describes a Goal Achievement strategy to improve countywide education where the core corrective actions involved strengthening relationships among diverse stakeholders and engaging in evidence-based continuous improvement.

BUILDING ON STRONG RELATIONSHIPS
TO IMPROVE REGIONAL FOOD AND FITNESS

When the W.K. Kellogg Foundation created their Food & Fitness program in 2006, there was growing public concern about childhood obesity. The program was designed to increase healthy eating and active living by all children, and the foundation had long supported developing a healthy, safe food supply and increasing consumption of good food. Because the issue was highly complex and prior efforts to address the issue in simpler ways had been unsuccessful, the lead program officers believed that a systems approach to Food & Fitness would increase the likelihood of:

- Engaging a diverse group of people and organizations.
- Fostering collaboration and finding innovative strategies to change the underlying systems.
- Creating and sustaining the healthy results everyone seeks for children and families.

The foundation brought together stakeholders from around the United States and also funded community systems change projects in nine geographic locations including northeast Iowa. Community leaders in this rural region developed the following vision for their initiative: "Northeast Iowa is a unique place where all residents and guests experience, celebrate, and promote healthy locally grown food with abundant opportunities for physical activity and play EVERY DAY. Healthier people make stronger families and vibrant communities."[7]

In order to realize the vision, they articulated a Success Amplification theory of change based on the historically strong relationships among people in the region. They believed that cultivating these relationships would help them move toward more collective thinking about how to take advantage of their agricultural base and open space, as well as more collaborative action, better results, and even better relationships. The virtuous cycle linking relationships, thinking, action, and results is a core theory of success originally developed by Daniel Kim.[8] Community leaders also recognized potential limits to growth by acknowledging that delays in learning and working across boundaries, as well as in converting innovative ideas into new policies, would try people's patience. To overcome these limits, they invested in collaborative technologies, engaged policy makers early in the process, and set realistic expectations around what could be accomplished in a given time frame.

Figure 11.9 summarizes their theory of change with particular emphasis on the engines of growth and reinforcing mental models.

Ann Mansfield, a co-convener of the initiative and now project director, summarized the initial value of using systems thinking from a grantee's perspective: "The tools helped us put a pause on the quick fix." By acknowledging the strengths they were building on, the time delays they expected, and the steps they planned to take to reduce delays where

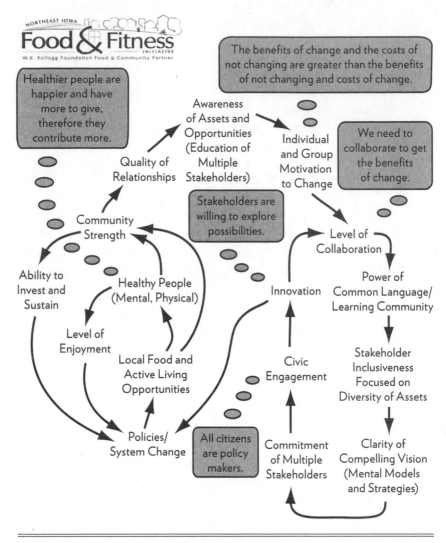

FIGURE 11.9 SUCCESS AMPLIFICATION FOR NORTHEAST IOWA FOOD & FITNESS INITIATIVE. This theory of change posits that building on the strong relationships that exist in the region increases collaboration, which in turn increases innovation, leading to policy change, healthier people, and even stronger communities. This dynamic is reinforced by holding certain assumptions. Northeast Iowa Food & Fitness Initiative

they could and increase people's patience where they could not, she and her colleagues established a robust and realistic way forward. Now, more than six years after creating the plan, Mansfield feels that "our core theory of success has, indeed, held up over time." She and the team use it continuously to stay focused on systems change. It helps them bring on new partners, orient new staff, check their progress, and create new strategies or lines of work.

CREATING A COMMUNITY WHERE
ALL CHILDREN ARE LOVED AND SUCCESSFUL

The InteGreat! coalition, committed to the health and success of all children in its highly diverse student population, also emphasized the importance of relationships in developing its strategy. Similar to the coalition formed by ECPHE, they chose to develop a Goal Achievement theory of change since their emphasis was on creating an inclusive community where all children are loved and successful in life. Moreover, because the population of the community was more diverse than in northeast Iowa, in terms of both income and ethnicity, they viewed improving relationships as a key part of the strategy. The coalition identified seven key success factors, of which four addressed relationship building directly:

- Quality of Community Relationships (among organizations impacting families).
- Level of Collaboration and Integration.
- Level of Youth Engagement.
- Level of Family Engagement
- Level of Data-Driven and Evidence-Based Practices.
- Access to Opportunities.
- Equity.

They developed a systemic theory of change—mapped out in figures 11.10 to 11.13—to link these factors and include others identified during the design conversation. In the initial map, figure 11.10, the core loop (B1) identifies the gap between actual and desired success as the driver of improving the quality of collaborative relationships, which in turn will improve the level of collaboration and integration among organizations impacting families, thereby creating mutual short-term successes, greater access to quality education opportunities, greater system equity, and a greater presence of educational success factors for all kids.

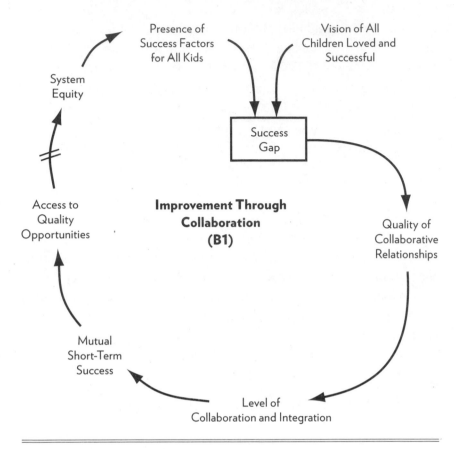

FIGURE 11.10 COLLABORATING FOR LOVED AND SUCCESSFUL CHILDREN. Improving collaboration among service providers is seen as the key to supporting all children to be valued and successful because it can lead to both mutual short-term successes and access to quality opportunities that increase system equity and the presence of success factors for all kids.

Building upon this map, figure 11.11 adds loops B2 and B3, which emphasize the benefits of higher-quality collaborative relationships to more effectively engage youth and families.

Figure 11.12 highlights the importance of strengthening data-driven and evidence-based practices. B4 focuses on developing shared knowledge. B5 and B6 show how agreeing on shared measures leads to more effective and efficient resource use as well as supporting mutual short-term success, both of which increase access to quality opportunities.

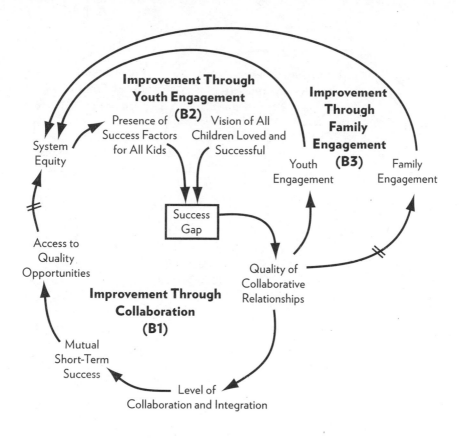

FIGURE 11.11 ENGAGING YOUTH AND FAMILIES FOR LOVED AND SUCCESSFUL CHILDREN. Improved collaboration among service providers also enables more effective outreach to youth and their families, thereby further increasing system equity.

The coalition also defined several reinforcing loops to ensure that initial improvements would be sustained and strengthened. R7 in figure 11.13 emphasizes the need to develop aligned infrastructure beginning with a backbone organization. R8 notes that increasing system equity can further support access to quality opportunities. R9 identifies the benefits of engaging youth to involve their parents as well as supporting parents to engage their kids. R10 explains that shared measures help to further focus data analysis.

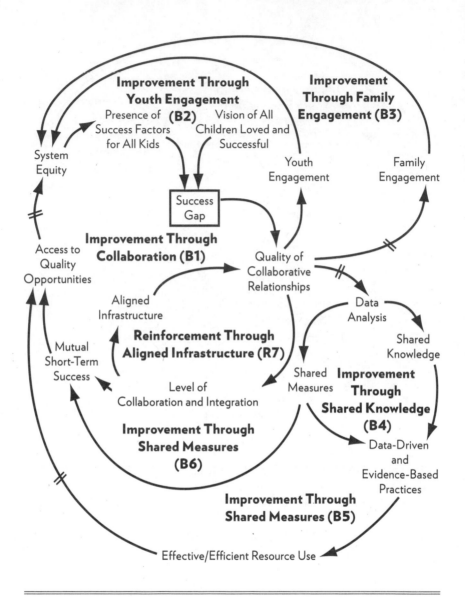

FIGURE 11.12 MANAGING DATA IN SUPPORT OF LOVED AND SUCCESSFUL CHIL-DREN. The development of more useful data leads to shared knowledge, shared measures, and improved data-driven and evidence-based practices. These in turn facilitate mutual short-term successes and the more effective and efficient use of resources, and both of these increase access to quality opportunities.

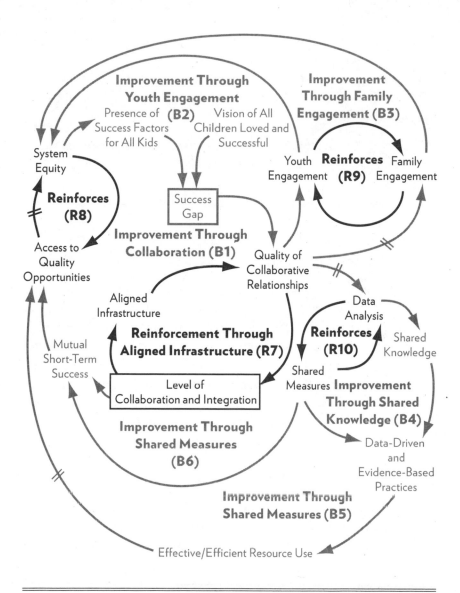

FIGURE 11.13 REINFORCING IMPROVEMENTS FOR LOVED AND SUCCESSFUL CHILDREN. Initial improvements can be reinforced in four different ways: capitalizing on increased collaboration and integration to develop an aligned infrastructure, further increasing access to quality opportunities through greater system equity, stimulating closer ties between youth and their families, and using shared measures to strengthen additional data analysis.

Often people distill more complex theories of change such as the one above into simpler insights over time. The InteGreat! coalition subsequently recognized that its theory was based on three key ideas. Traci Wodlinger, the chief strategy officer for the Eagle County Schools, summarized them as:

1. A collaborative trust that aligns currently existing organizations and their assets will improve the equity of opportunities for our children.
2. Families and children must be engaged in the work, both directly and indirectly.
3. We must utilize continuous process improvements, where data will drive change and maximize efficiency of our efforts and resources.

Note that the change theory depicted in these diagrams includes all of the elements required for Collective Impact as defined in chapter 2. The systemic theory of change *is* the common agenda and explicitly incorporates shared measures and the need for a backbone organization. It also provides a road map for mutually reinforcing activities and continuous communication. Moreover, these critical success factors as well as others emerged directly from conversations among community leaders.

The consultant to the overall project, Kathleen Zurcher, reports several benefits resulting from the process of designing this systemic theory of change:

- The map represents the coalition's thinking about how they want to move forward, and they refer to it frequently as their road map.
- The smaller steering committee and larger group that participated in this process are translating the loops into narratives to explain the plan to their respective constituents.
- Developing the map increased people's understanding of inequities in the community and stimulated their first honest and open talk about the issue.
- They now have a common set of definitions going forward.
- The map helps the coalition focus on systemic change, assuring that strategies, tactics, and activities are aligned and contribute to intended outcomes.

Streamlining Choices

The following case describes the development of a systemic theory of change to organize an agency's large numbers of programs and tasks into a manageable strategy. By distilling a coherent strategy from what often seemed like disparate elements, the organization was able to focus its commitments and avoid win–lose debates about the relative importance of different people's work, which often make priority setting so difficult.

A large child welfare agency was concerned that it was spread too thin. The budget was too low to sustain all of its programs at a high level of effectiveness. Moreover, there was competition for funding between program management and the agency's other functions, in particular research & evaluation, and advocacy. When time came to review the agency's strategic plan, the senior management team decided to develop a systemic theory of change. By making better sense of the connections among the organization's various efforts, the team hoped to streamline programs and reduce tensions between functions. The team first categorized its programs—more than thirty—into four primary categories:

- **Prevention:** Helping families at risk develop and maintain the stability required to provide a safe, nurturing home for their children.
- **Stabilization:** Stabilizing children who have to be removed from their homes in transitional homes managed by the agency.
- **Development:** Supporting the nurturance and continued education of children in these homes.
- **Placement:** Reunifying children with their biological families or providing alternative permanent placements for them in the community.

The team then articulated the links across these categories. If prevention failed to help families overcome the many external pressures they faced, resulting in an unsafe and unsupportive home environment, children were removed from their home by law and placed by the agency in the transitional homes it managed. The goal at this point became to return the children to a safe, nurturing, and permanent home by creating a core path to recovery that provided stability followed by development and finally reunification or alternative placement—as shown in figure 11.14, loop B1.

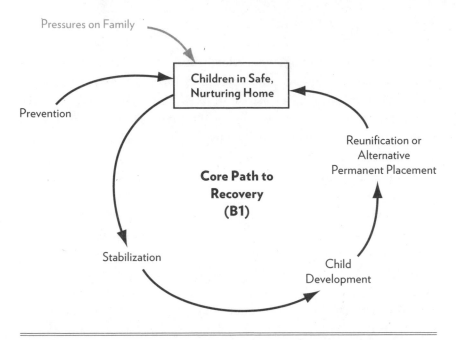

FIGURE 11.14 CREATING A SAFE AND NURTURING HOME FOR CHILDREN. When pressures on the family become too great, children must be temporarily removed from the home and sustained through first stabilizing them and then facilitating their ongoing development. Providing this foundation outside the home gives children a strong base for reunification or alternative permanent placement.

These four program categories required ongoing resources to sustain them. The management team clarified how other functions needed to be integrated to ensure the agency's financial viability. The agency maintained a strong research and evaluation unit. However, many program managers believed that the unit not only drew money away from their vital work on the ground, but also failed to provide fair evaluations of their work. The team was able to identify that, when the unit conducted assessments according to clear and agreed-upon criteria, it provided invaluable information to the agency that not only strengthened programs but also created additional benefits. One immediate benefit was to directly increase fund-raising effectiveness, shown by R2 in figure 11.15, and a second was to demonstrate evidence-based practices that could be used to advocate for changes in child welfare policy. More effective advocacy would further increase fund-raising effectiveness (R3), and the

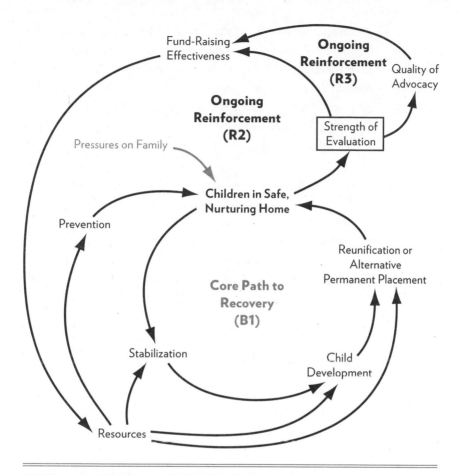

FIGURE 11.15 ENSURING SUSTAINABLE RESOURCES OVER TIME. Evaluation and advocacy are key to the effective fund-raising required to attract additional resources in support of the agency's core programs.

increases in fund-raising effectiveness would provide ongoing resources for the agency's program work.

As a result of articulating its systemic theory of change, the agency was able to identify and eliminate programs that did not fit into its strategy. Interestingly, the managers of these programs had often taken up the most airtime in defense of their work, perhaps because they and their colleagues sensed that the programs were not a good fit for the organization. These programs were either transferred to other community organizations better suited to manage them or phased out when current funding ended.

The theory also enabled all of the functions to better appreciate and integrate their diverse contributions to ensuring children lived in safe, nurturing homes. For example, the evaluation and advocacy units resolved tensions about their relative importance by recognizing that the organization needed to invest more in evaluation in the short run to increase the quality of its advocacy in the long run. Often, *sequencing* is an essential aspect of prioritizing because it enables people to think about the relative importance of different work over time.[9] By contrast, prioritizing often results in win–lose debates that drain organizational energy and goodwill because people assume that, if their work is not considered a high priority now, it never will be. By taking time delay into account, systemic theories of change circumvent such debates and enable many parties to affirm their contributions over time. In the case of the evaluation and advocacy functions, the agency agreed that it needed to step up investment in evaluation in the short run to strengthen its platform for evidence-based practices, and shift more funding to advocacy over time as this platform became more established.

How to Refine Your Systemic Theory of Change

Systemic theories of change must be designed to evolve over time as more information becomes available, actions take effect, and conditions shift. Refining your theory means incorporating the theories of other stakeholders, tracking how key elements change over time, and modifying the theory by comparing what you expect to happen with what actually occurs.

In all of the cases discussed in this chapter that involved diverse community stakeholders, a larger group of constituents was invited to refine the theory developed by a smaller leadership group. It is important to ensure that refinements at this stage come not only from people within the organizations represented by the leadership group, but also from other key stakeholders who might not have been involved initially. These often include the ultimate beneficiaries of the work (such as students, patients, and homeless people) and the private sector, which is often misperceived by core participants to have less of a stake in social issues or be less accessible. Integrating these diverse views often presents the same challenges as when you develop a systems map to deepen your understanding of why a chronic problem persists: The systemic theory must be rich enough to incorporate multiple perspectives yet simple enough to provide actionable insights.

Tracking your theory involves identifying key indicators for each critical variable, estimating time delays, and drawing a pattern of behavior that projects how each key success factor or performance measure should change over time. Another element to track is mental models: identifying how people's assumptions need to evolve over time and estimating (often with long time delays) when to expect changes to become evident.

Because learning takes time and can lead to costly mistakes, it is important to establish milestones along the way to test more quickly if you are on or off course. In addition, by developing a computer simulation of your theory, you can shorten time delays in the real world to a matter of seconds and thereby accelerate experimentation and learning. For example, computer simulations exist for addressing climate change and sustainability and for testing alternative models of health care delivery.[10] While computer simulations require more up-front investment, they can quantify outcomes and reduce the learning curve for testing different theories of change.

By comparing the intended and projected behavior over time with what actually occurs, you engage in continuous learning and can refine your assumptions and actions. You are able to rethink important cause–effect relationships, alter your strategy for revising mental models, and/or adjust expected time frames. Tracking and modifying your theory of change will also be examined more closely in the next chapter, on evaluation.

Closing the Loop

- Circular theories of change create pictures that quickly communicate a lot of readily understandable and navigable information.
- Applying systems thinking to strategic planning can help organizations and communities organize the leverage points identified by a root cause analysis, integrate the multiple critical success factors required to create something new, and streamline choices among too many programs and priorities.
- Based on the tools used in this book, there are two core theories of systemic change: one that seeks to amplify success and the other that strives to improve an existing situation.
- Amplifying success includes anticipating limits to success and planning ways to overcome these limits.

- Achieving goals includes planning to sustain and build on progress over time.
- Sequencing strategies over time can break the win–lose dynamics often associated with prioritizing.
- Systemic theories of change need to evolve—and this can be accomplished by incorporating the theories of other stakeholders, tracking what actually happens against your projections of what should happen, and modifying your theory based on analyzing and seeking to bridge this gap.

CHAPTER 12

Systems Thinking
for Evaluation

Funders want to determine whether or not their investments are making a positive difference, and recipients want to show that they are meeting their objectives so that they can continue to obtain funding. However, evaluating solutions to social problems is not easy. One of the challenges is the complexity of the problems themselves. In addition, both funders and grantees tend to set unrealistic goals—the former because of their desire to see results quickly, and the latter in part because of their high aspirations and in part because of their desire to impress funders. Policy makers in the public sector sometimes legislate unfunded mandates, making missions difficult to fulfill. Grantees tend to resist monitoring and evaluation efforts because they do not want to set aside limited funds for this work and are concerned about uncovering negative evaluations. And goals change and priorities shift over time—making it difficult to establish the end points against which performance is being measured.

Still, evaluation remains critical, and private and public funders often ask how systems thinking can help contribute to the process. As we saw in chapter 11, systems thinking contributes to evaluation in part by helping people articulate a systemic theory of change with key indicators, metrics that can be tracked over time, and predicted patterns of behavior for the indicators that can subsequently be compared with actual results. Now we'll look at the general systemic guidelines for improving evaluation—including ways to meet the challenges described above—and learn how to assess progress against the two theories of change also introduced in chapter 11.

General Systemic Guidelines

You can use systems thinking to inform your evaluation process in five ways:

- Set realistic goals.
- Define clear key indicators and metrics.
- Think differently about the short and long term.
- Look for consequences along multiple dimensions.
- Commit to continuous learning.

SET REALISTIC GOALS

Unrealistic goals hurt both the party setting the expectations and the party agreeing to meet them. It is a recipe for toxic stress, broken agreements, and frayed relationships. This is not to discourage the power of an aspirational vision. As Henry David Thoreau said, "If you have built castles in the air, your work need not be lost; that is where they should be. Now put the foundations under them."

Goals are part of the foundation; they provide milestones to strive for toward realizing the vision. It helps to set a limited number of goals at any one time; no more than three is a recommended rule of thumb.[1] In addition, a systemic theory of change helps people define the scope of their intentions, suggests an order to follow in achieving them, and indicates time delays to address along the way. A good theory enables people to distinguish short- and long-term goals and discourages them from trying to do everything at once.

While people might be deliberate in setting a limited number of realistic goals at the outset, it can be difficult to sustain focus on these goals over time. New conditions and priorities emerge that can easily overwhelm us or take us off course. The challenge is to shift priorities consciously if at all, and to slow down or stop work on former commitments in the face of limited resources. By contrast, taking on new work without letting go of other work leads people to become unproductively stressed and reactive instead of strategic. How people react to overload tends to create vicious cycles that increase workload further over time. To avoid getting caught in these cycles, it helps to deliberately and regularly rebalance workload with available resources.[2]

DEFINE CLEAR INDICATORS AND METRICS

As noted in chapter 10, it is important to focus on, and if necessary create, indicators and metrics that are aligned with people's chosen purpose. In selecting them, you can include both qualitative and quantitative measures—both of which are important in developing rich systems maps. For example, both the level of collaboration among service providers and biometric levels of health are important factors to track in building a healthy community. It can help to break down a more abstract variable such as collaboration into elements that may be more measurable, such as the use of a common language and commitment to shared goals and strategies. Interviews with key stakeholders and opinion surveys are two ways of gathering qualitative data, which also includes the mental models held by different stakeholders.

One of the benefits of systems thinking is that it helps people look for leverage. One way to think about leverage is that some actions produce a higher return on investment (of time, money, people, or other inputs) than others. The concept of measuring returns through ratios instead of absolute numbers was critical to three types of metrics developed by the American Productivity and Quality Center, which distinguished measures of productivity, effectiveness, and efficiency.[3] The center defined productivity as the ratio of outcomes to inputs, where outcomes are the results achieved by intended beneficiaries and inputs are the resources expended. Effectiveness was the ratio of outcomes to outputs, where outputs are what the resources produce. And efficiency was the ratio of outputs to inputs.

For example, it is not enough to know that a jobs program costs fifty thousand dollars (an input), provides one hundred hours of training (an output), or even results in 70 percent of the graduates getting living wage jobs and holding these jobs for at least one year (an outcome). It is important to also evaluate the number of such quality jobs created per dollar spent on the training (productivity), the number of quality jobs created per training hour (effectiveness), and the number of training hours per dollar spent (efficiency).

When my colleague John McGah and I were asked by a statewide homeless coalition to propose meaningful indicators for ending homelessness, we proposed such systems-thinking-based metrics as:

- Overall Effectiveness = (# homeless people per month moved into permanent housing) – (# people per month becoming homeless): *Look for monthly decline in # homeless people.*
- Prevention Effectiveness = (# people per month becoming homeless) ÷ (# people per month at risk of becoming homeless due to income level and family stressors): *Look for decline.*
- Living Wage Job Creation Dividend = (rent paid by tenants + income taxes generated per tenant per year) – ($ spent per person per year on job training and relocation): *Suggests annual $ to reinvest in job creation.*

THINK DIFFERENTLY
ABOUT THE SHORT AND LONG TERM

As noted in chapter 3, distinguishing between quick fixes and small successes enables people to increase the likelihood of accurately evaluating short-term improvements. Quick fixes are solutions that work in the short run, tend to produce longer-term consequences that neutralize or reverse the initial outcomes, and are usually developed without a clear understanding of the root causes of the problem or a long-term strategy. By contrast, small successes are short-term results that build momentum toward a long-term strategy based on a deep appreciation of the current issue.

There are six things you can do to determine if the short-term gains achieved by an action are also likely to lead to long-term improvements:

- Remember that systems tend to exhibit better-before-worse behavior.
- Refer to the relevant systems analysis that distinguished problem symptoms from root causes and ensure that your improvement addresses root causes.
- Inquire into the systemic theory of change that informed the choice of this action.
- Ensure that time delays were considered in arriving at the actions to be taken, and that people either accept these delays or identify strategic ways to shorten them.
- Look for early successes that build systemwide capacities (in relationships, insight, and organization infrastructure) instead of

those that immediately reduce problem symptoms such as low test
scores or poor health.

- Build a computer simulation of the issue to quickly test the differences
  between the short- and long-term consequences of various actions.

LOOK FOR CONSEQUENCES
ALONG MULTIPLE DIMENSIONS

In addition to evaluating impacts differently in the short and long term, it
helps to look for consequences along other dimensions. First, some consequences are intended while others are not—and both provide valuable
information and implications for next steps. Second, not all unintended
consequences are negative, and it is important to track and build on positive unintended consequences as well. Third, it is important to look for
resources that are developed and/or saved as well as those that are used to
get a complete picture of the return on your investment.

An excellent video describing all of these impacts relates the surprising
story of efforts to bring fresh potable water to people in rural Togo, West
Africa, by building wells in local villages.[4] The wells built by government
engineers were initially met by positive responses from villagers because
they replaced their traditional water supply obtained from water holes
infected by guinea worm, a painful and sometimes life-threatening parasite,
and fetched by village women making long walks often two or three times
a day. However, after two years the wells were abandoned because the main
part of the pump broke, and neither the villages nor the government had
developed the infrastructure to maintain them.

In the case of Togo, the engineers went back to the villages and learned
why the wells had fallen into disuse. They realized that they needed to help
the villagers learn to do their own maintenance. This involved working with
villagers to develop the local infrastructure and skills required to identify
the causes of breakdowns, collect money to repair the pump, and fix the
pump. The government also stocked parts for the pumps in nearby towns
to ensure that villagers could make their own repairs. The infrastructure
created by the villagers enabled them to not only maintain their wells but
also initiate and manage their own development projects that provided
them with additional food and income.

COMMIT TO CONTINUOUS LEARNING

Because systems are complex and evolve in unpredictable and often unexpected ways, evaluation must be approached as an ongoing activity. You can facilitate continuous learning in several ways:

- If you are a funder, commit to longer-term investments. This makes it easier for both funders and their beneficiaries to learn from failures as well as successes since people no longer feel compelled to downplay or hide the former and inflate the latter.

- Continue to engage existing stakeholders and involve new ones. Active stakeholder engagement not only provides the feedback essential to learning and adapting over time, but also builds increasing levels of ownership among more people.

- Use feedback from experiments and stakeholder involvement to refine your systems analysis and theory of change over time. The more accurate your understanding of what is happening and why, and the clearer your road map going forward, the more effective you will be.

Tracking Success Amplification

Because a Success Amplification theory of change has a recognizable trajectory, you can look for more specific indicators to evaluate if you are on track in implementing it. The S-curve in figure 4.6—replicated in a somewhat different format in figure 12.1—is the pattern of natural organic growth. It moves through three phases: an initial phase of slow growth where progress is not readily visible, a second phase of steep growth where progress is dramatic, and a third phase of maturation where growth slows down and plateaus. The challenge for most people, including policy makers, funders, and managers, is that their expectations for linear growth are out of sync with this pattern, as shown in figure 12.1. We tend to expect more measurable progress than is natural in the early phase, can become overwhelmed by rapid growth in the second phase, and become complacent when progress exceeds our expectations in the third phase. People who expect even faster improvements in the short run are likely to become even more frustrated than those with linear expectations.

The key for evaluators is to know what to expect and look for in each phase of the growth process. The important changes to cultivate and assess in the first phase have to do with foundation building: developing the common ground, systemwide relationships, and organizational capacities required for innovation and the management of growth. Improvements in these areas constitute the small successes that build momentum for long-term success.

Establishing common ground means defining people's shared aspirations in terms of their vision, mission, and values. It also entails developing a common understanding of current reality by beginning to progressively fill in the iceberg introduced in chapter 3, and eventually agreeing on a shared systemic theory of change as described in chapter 11.

Relationship building needs to be assessed in three areas: across service providers who would otherwise tend to try to optimize just their part of the system, across institutions including not only the social and public

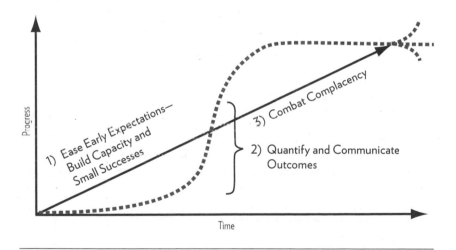

FIGURE 12.1 ALIGN EXPECTATIONS WITH ORGANIC GROWTH. People's expectations for linear growth are not in alignment with the S-pattern of organic growth. You can align linear expectations with organic growth by developing capacity and small successes in the short term, quantifying and communicating outcomes in the medium term, and transforming complacency into innovation in the long term. Innovation fuels a new engine of growth leading to an upturn in performance; however, the failure to innovate at this point results in a flattening or downturn in performance over time.

but also business sectors, and between these groups and the intended beneficiaries of change—the people whose voices are often not heard or sufficiently appreciated. In this phase, an evaluator should ask: Are there forums for stakeholders to come together and make new connections? Are those involved developing the collaborative capacities of systems thinking, partaking in productive conversations, and taking responsibility for the performance of the whole system as well as for their part?

It is also vital to build organizational capacity, especially in nonprofits where the expectation has been that overhead should be as low as possible in relation to program delivery. In their section "Reforming the Art of Helping" in *A Path Appears*, Nicholas Kristof and Sheryl WuDunn point out that underinvesting in the infrastructure and business skills of non-profits significantly reduces their impact. It is important to help them be more effective by investing in such organizational capacities as computers and information systems, marketing and customer relations, personnel development and talent, and evaluation.[5]

Developing the above foundation in phase one positions the system to more effectively manage the rapid growth to be expected in the second phase. Here is where it makes sense to measure outcomes and compare them to the outputs generated and inputs invested. Here is where it helps to gauge the benefits of saved and more productive lives against the costs of doing nothing.

The third stage begins with a certain amount of leveling off or maturation. The key at this stage is to shift from a tendency toward complacency with past successes to a commitment to innovation and finding new ways to grow. Although encountering limits to growth as identified in chapter 4 is inevitable, it is also possible and essential to overcome these limits by identifying new engines of growth at this stage instead of simply pushing harder on the old ones. Success in making this shift is measured in terms of sowing seeds for a different approach to development instead of allowing performance to level off or gradually decline.

Tracking Goal Achievement

A Goal Achievement theory also has a recognizable trajectory. It has two characteristic phases: correction (or improvement) and reinforcement.

Evaluators can pay attention to several indicators for the likely success of each phase.

Many of the indicators of success in phase one of Success Amplification are similar to the ones for correction, while others require special attention here. Similarities include establishing common ground, building relationships and relational skills, and developing organizational capacities. What is distinct is the often additional investment required to describe current reality. It is especially important in determining corrective actions to first clarify the root causes of the complex, chronic problem that people are trying to solve, or to help answer the question, "Why haven't we been able to achieve this goal despite our best efforts?" Equally important is to recalibrate this understanding if people are not progressing as expected.

As demonstrable progress is being made, it is vital to ensure that people maintain momentum through conscious reinforcement. This means ensuring that goals increase along with actual performance, that people identify and invest in new growing actions to achieve these goals, and that dividends from initial improvements are redirected to fund these new actions.

The bottom line of thinking systemically about evaluation is that assessment involves much more than measuring outcomes. It includes realistic goal setting; clarifying a range of indicators and metrics aligned with the chosen purpose, both quantitative and qualitative; thinking differently about the short- and long-term impacts of actions; looking for consequences along multiple dimensions; and committing to continuous learning. Success Amplification and Goal Achievement theories of change share certain common indicators of success in their initial phases, but vary in terms of what evaluators need to look for to measure success over time.

Closing the Loop

- Complexity is only one of several challenges to evaluation that systems thinking can help you meet.
- A clear systemic theory of change provides a strong foundation for effective evaluation.
- To increase the likelihood of positive evaluations, ensure that organizations set realistic goals in the first place and consciously rebalance workload with resources over time.
- Track both quantitative and qualitative metrics.

- Use six guidelines to distinguish quick fixes from short-term successes.
- Look for consequences that are unintended as well as intended, negative as well as positive, and that create as well as expend resources.
- Support continuous learning.
- Evaluate progress toward amplifying current successes by distinguishing three different phases of growth.
- Evaluate progress toward bridging the gap between current reality and system goals by distinguishing two different phases of improvement.

CHAPTER 13

Becoming a Systems Thinker

Becoming a deeply skilled systems thinker takes time, but it is definitely possible. We've learned that, on one level, systems thinking is child's play: We are born with the capacity to see connections and understand (though not necessarily tolerate) time delay. We've also learned that the very work of applying systems thinking tools and practices not only hones that capacity but also shapes who we are and how we see the world. This orientation in turn increases our effectiveness in applying the tools and practices.

This final chapter focuses on three ways you can become a more effective systems thinker. First, it helps you develop what I would call a systems orientation. Since we are a part of the system we seek to understand and influence, we need to cultivate a certain way of being and also learn new ways of doing things. We need to appreciate what Peter Senge calls the *essence* of the discipline—the state of being or life orientation that infuses your work as a systems thinker.[1] Thinking systemically is paradoxically not just a mental discipline but also an emotional, physical, and ultimately spiritual one.

Second, the chapter describes an effective way of learning to apply systems thinking for social change. It introduces a facilitated action learning approach that enables you to integrate systems analysis tools into the change management framework of this book and guides you to apply the framework and tools to a real change challenge in your organization or community. Finally, the chapter reviews a set of key questions you can ask to open the door for others to think systemically—whether or not they are visual thinkers, can draw a systems map themselves, or have previously learned to read these maps.

Develop a Systems Orientation

Donella Meadows observed, "Social systems are the external manifestations of cultural thinking patterns and of profound human needs, emotions, strengths, and weaknesses."[2] In other words, they cannot simply be changed by dictate, or even by introducing a few good ideas. Our effectiveness as social change agents is impacted by our own capacities and character traits as well as by the multiple dimensions of human experience that we seek to influence.

It is tempting to view systems thinking as just a mental discipline. However, this perspective misses the richness and breadth of the approach—which also includes emotional, physical, and spiritual dimensions. Integrating all of these dimensions increases your effectiveness in applying systems thinking to meet the complex challenges that organizations and other social systems face. Let's look at each of these four domains in turn.

MENTAL

Systems thinking is a language and set of tools meant to illuminate our thinking about how the systems we are all part of actually operate. Built into this language are important core principles about how systems function:

- **Feedback:** Performance of our organizations and systems is largely determined by a web of interconnected circular (not linear) relationships.
- **Growth and Stability:** Feedback helps us understand how systems both grow and remain stable.
- **Diversity and Resilience:** Systems grow and innovate through diversity, and they remain stable because of their resilience in the face of change.
- **Delay:** Actions we take have both immediate and delayed consequences that we do not always consider.
- **Unintended Consequences:** Today's problems were most likely yesterday's solutions.
- **Power of Awareness:** When we see and understand a system as it really operates, we build on its inherent strengths and avoid being controlled by its weaknesses.

- **Leverage:** Systems improve as the result of a few key coordinated changes sustained over time.

Systems language and principles often combine to create recognizable patterns or classic stories. The core story describes how people tend to fall short of the results they want despite their best efforts. System archetypes are all variations on this theme. They describe different ways in which these shortfalls can result from people's underlying assumptions and actions, as well as from the multiple and sometimes conflicting goals that compete for their attention.

EMOTIONAL

It can be very difficult to acknowledge how your own thinking and behavior contribute to the problems you want to solve. One tendency we have is to blame others instead of taking responsibility when things do not work out as planned. Systems thinking is one of the tools we have to help us shift from blame to responsibility and hence empower ourselves by taking charge of our own reality.[3] A related tendency is to try to compensate for feelings of powerlessness by trying to control the system instead of partner with it. When we empower ourselves we are more able to work with the forces at play in the system—to both appreciate what works about it and respectfully address what does not—instead of to unconsciously work against the forces and often make things worse.

No matter how limiting our beliefs and assumptions, we tend to be emotionally attached to them because we equate who we are with what we think. Moreover, we are frequently rewarded for strongly advocating our beliefs. Therefore, changing how we think involves the humility, curiosity, and courage to take such emotional risks as admitting we might not be right, experimenting with new assumptions and behaviors, and learning from others. It is important to be accepting of everyone's views since they can contribute to our own understanding, and to be compassionate toward them since all of us have our own limited perspectives.[4] These are the competencies of emotional maturity and social intelligence.

PHYSICAL

Systems thinking is a team sport. It works because stakeholders with diverse perspectives come together to share their views, expand their understanding, and develop a more complete picture of the reality they all face. The primary purpose of mapping a system is to stimulate catalytic conversations that lead to shared insights and shared responsibility, which in turn provide the foundation for *coordinated action*.

Coordinated action takes place in the physical realm. It is made possible through a combination of *convening* systemically—bringing diverse stakeholders together to share their aspirations, viewpoints, and experiences—with *thinking* systemically—understanding a complex problem in terms of the interdependence of its parts. Convening systemically by itself tends to encourage people to try to optimize their part of the system at the expense of the whole, and thinking systemically in isolation tends to produce insights that people do not identify with or want to support.

SPIRITUAL

Systems thinking is a spiritual practice because it helps you see that:

- Everything is connected.
- You have choices about furthering positive connections—or feeding dysfunctional ones.
- In order to make constructive choices, it helps to develop certain character traits.

Learning to see connections, make good choices, and cultivate character strengths are all spiritual practices.

See Connections

Many spiritual traditions are based on a belief that everything is connected. The three monotheistic Western religions stem from a belief that we all come from and hence are connected by the same life source. In Buddhism, Indra's Net symbolizes a universe where infinitely repeated mutual relations exist among all its members. In Hinduism, the true reality is the inner spiritual principle Atman-Brahman, which gives life and being to all things.

We hurt ourselves and the world around us when we fail to recognize and cultivate our essential connectedness. The very word *religion* has its roots in the Latin word *religare*, commonly translated as "to bind." It is all about making connections. From this viewpoint, systems thinking can be viewed as the work of enabling people to make connections in service of the whole. Serving the whole has both moral and practical benefits: We strive to contribute to the greater good, including our own, and to build stronger support for change by appreciating everyone's interests. The connections we generate are not only emotional ones among people but also logical ones among the parts of the system that people identify with. Systems thinking enables us to transform the parts of a more complex problem into a shared understanding of the larger issue, and to organize parts of a strategy into a clear direction and navigable road map.

Make Good Choices

Just because everything is connected does not mean that all connections are positive. We can be connected for better—such as when my esteem of you enhances yours of me—or for worse, as when my disregard of you increases yours of me. In a different context, the market dynamics that fueled the housing boom also created the deep recession of 2008.

Connecting well means:

- Orienting your actions toward goals that serve the whole over time.
- Optimizing relationships among the parts of the system—instead of seeking to optimize just your part.
- Clarifying and expanding the boundary of the system for which you feel responsible.

Clarifying goals, cultivating positive relationships, and defining system boundaries are all choices. One powerful example is the story of pediatric surgeon Victor Garcia, the founder of Trauma Services at Cincinnati Children's Hospital. Garcia's patients included young children who had been caught in the crossfire of urban violence. Despite his best efforts as a doctor, he was not able to save the lives of all of them. One day the parents of a young boy who had died in his care sought consolation by asking him if he had done the best he could to save their child. In that moment, he realized that the answer was no. By confining his work inside the walls of the hospital, he had isolated himself from the causes of violence that had led to the boy's death. As a result, he chose to redefine "doing his best" as reaching beyond the hospital to address the dangerous and unhealthy conditions of the inner city where his patients lived. He found CoreChange to develop systemic solutions to poverty and its effects on these children.[5]

Cultivate Character Strengths

In order to make choices that produce positive connections, it is important to develop certain qualities or character traits within ourselves. These include:

- **Curiosity:** Being open to learning, particularly in the face of failing to achieve what you really care about.
- **Respect:** Assuming that everyone is doing the best they can with what they know at the time.
- **Compassion:** Recognizing that at some level people are unaware of the harm they do and that limited awareness contributes to suffering.
- **Awareness:** Knowing yourself, seeing more of the whole of which you are a part, and understanding how you might unwittingly be contributing to the very situation you want to change.
- **Vision:** Listening for what moves you and what is being called for by the world around you. Working for what you deeply care about and remembering, in the words of Václav Havel, that "hope . . . is the certainty that something makes sense no matter how it turns out."
- **Courage:** Taking a stand for the integrity and sustainability of the whole in the face of seemingly more expedient alternatives. Going even farther and asking, "What might I or we have to give up in order for the whole system to succeed?"
- **Patience:** Developing the patience and persistence to stay the course in the face of uncertainty and time delay.
- **Flexibility:** Balancing the ability to stay on course with the flexibility to adjust in the face of new information.

Learn by Doing

Trying to solve chronic, complex problems is daunting, and you don't have to be an expert to take the first step. Your abilities as a systems thinker will grow over time, and you can learn by doing.

Often groups recruit help. Michael Goodman and I developed an approach that enables participants to immediately apply the principles and tools laid out in this book to achieve significant changes in their own organizations and communities. We follow this approach when working with a wide range of stakeholders—foundations, NGOs, public agencies, and private businesses. The process can be designed to: solve one or more problems, develop a systemic theory of change, and/or build capacity in systems thinking. People benefit from a combination of working meetings, training, and coaching in real time over an average three- to six-month period.

Whether or not you are working with a consultant or coach to organize your systems-thinking process, you may find it useful to introduce the basic language, principles, and tools of systems thinking to people in organizations and communities who can benefit from the approach but do not want to become proficient in it themselves. For those who prefer to learn online, Michael has developed an interactive web-based course called "Applying Systems Thinking and Common Archetypes to Organizational Issues."[6] There is also a wide range of additional resources on systems thinking that you can turn to (see Appendix D).

Finally, I encourage you to practice systems mapping on your own by taking any chronic, complex problem you are interested in—whether it be climate change, the expanding influence of money in politics, or a rift between you and a loved one—and using the tools outlined in chapters 3, 4, and 7 to understand it better. Working with an issue you are passionate about helps you overcome the initial awkwardness of learning something new and experience the satisfaction that can come with deeper insight.

Like any new language, systems thinking takes practice. I hope that the cases in this book inspire you to put in this practice so you can learn over time to create similar results for yourself with the people and causes you care about.

Ask Systemic Questions

One of the most effective ways to become a systems thinker and help others do the same is to ask powerful questions. You can do this whether or not you are a visual thinker, draw a systems map, or expect others to read these maps. It can help to remember that, while the maps are great prompts to create catalytic conversations, asking good questions also opens the door to new ways of thinking, communicating, and understanding.

By way of review, here are some of the most useful questions you can ask:

- Where do our best intentions fall short of achieving what we really care about?

- Why are we not as successful as we want to be despite our best efforts?
- What might be our responsibility for the obstacles we encounter and shortfalls we experience?
- Are there people who share similar aspirations to ours but have very different views about the nature of the problem and/or the solution? If so, what can we do to help align our respective efforts more effectively?
- What can we learn from a preliminary inquiry into specific events related to our issue, underlying trends or patterns of behavior over time, and a consideration of deeper systems structure?
- How might the concepts of time delay, archetypes, and the Bathtub increase our understanding of systems structure related to the issue?
- Which stakeholders are we comfortable engaging now, and what are their motivations for change?
- By contrast, which stakeholders might we not choose to engage at the outset—and why? What might we miss by not involving them initially, and what strategies do we have for engaging them over time?
- How can we create common ground among the stakeholders we engage now?
- How do we increase people's understanding of the issue in a way that integrates the richness of diverse perspectives with the simplicity required to act?
- How do we build support for an analysis that might be difficult to communicate or that challenges people's underlying beliefs and assumptions?
- What is the case for the status quo?
- What might we have to give up in order for the whole to succeed?
- What interventions could enable us to achieve sustainable, breakthrough change?
- What might be the unintended consequences of our proposed solutions?
- How do we ensure continuous learning and outreach?
- What is our systemic theory of change?
- How do we evaluate progress toward our vision using a system lens?

- What actions can we take to become better systems thinkers?
- What do we intend to do next?

A US Park Services manager observed, "I used to think of the organization as a machine, and that things like rockslides and traffic jams caused breakdowns in the machine; and now I see the organization as an organism where those are just events and the true sources of breakdown are egos, mental models, and poor communications/relationships." Becoming a more effective systems thinker includes developing an orientation and way of being that incorporates the emotional, behavioral, and spiritual—as well as cognitive—dimensions of life. It takes time and practice with real-world challenges. And remember, even if you never draw a systems map on your own, you can always ask powerful questions that open the door for you and others to think systemically.

Closing the Loop

- Systems thinking is not (just) what you think.
- Becoming a more effective systems thinker means devel oping your emotional, behavioral, and spiritual—as well as cognitive—capacities.
- The best way to learn is by doing, and there are many resources available to help.
- When you are not sure what to do next, ask a systemic question.

ACKNOWLEDGMENTS

The ideas that have gone into writing this book stem from the contributions of many people over nearly forty years.

I first want to thank my long-time colleague Michael Goodman, a leading systems thinker who shares my passion for enabling leaders and organizations to apply these principles and tools to real-world issues in easily accessible ways. Michael and I co-created the change-management framework that appears in chapters 5–10 and the Leading Systemic Change workshop, which we have conducted for nearly twenty years in a wide range of organizations, leadership-development institutes, and professional conferences. As the director of the systems-thinking practice at Innovation Associates (now Innovation Associates Organizational Learning), Michael also contributed to the development of the systems archetypes and many of the other tools used in this book and by a generation of systems-thinking practitioners. He was also kind enough to read the entire manuscript and offer valuable improvements.

I also want to especially thank two other colleagues with whom I have worked closely to develop and test the ideas in this book. John McGah, the founder of the national nonprofit Give US Your Poor, strongly reinforced the benefits of using systems thinking to motivate diverse stakeholders to transcend their immediate self-interests and apply best practices to ending homelessness, particularly when those practices challenge more traditional ways of thinking.

Kathleen Zurcher has inspired and given me many opportunities to help communities develop systemic theories of change. As the former chief learning officer at the W.K. Kellogg Foundation, she has also collaborated with me in developing ways to explain the value of applied systems thinking to the work of foundations.

I am very grateful to my fellow co-founders of Innovation Associates, who gave me the opportunity to cultivate and promote the disciplines of organizational learning. Charles Kiefer was the primary founder of Innovation Associates and helped pioneer the field through his commitment to creating organizations as vehicles for people to realize their highest aspirations. He orchestrated the integration of traditional organizational development, systems thinking, and personal mastery into a compelling synthesis, which fellow cofounder Peter Senge subsequently expanded on to write the management classic *The Fifth Discipline* and later books that popularized systems thinking and its connection with the other learning disciplines. Robert Fritz was also a cofounder of Innovation Associates and a master of the creative process, which imbues all of this work with principles and tools for enriching the lives of individuals, organizations, and the people they serve.

I also want to thank two other colleagues at Innovation Associates for their contributions. Jennifer Kemeny, a brilliant systems thinker, helped me develop the original Leading Systemic Change workshop along with Michael Goodman. Sherry Immediato opened the doors for applying systems thinking to the issues of public and environmental public health.

Other colleagues have also played important roles in the development of this book. Joe Laur and Sara Schley invited me to collaborate with them on The After Prison Initiative, Curtis Ogden invited me to partner with him on Right from the Start, and Daniel Kim introduced Michael Goodman and me to the W.K. Kellogg Foundation, which sponsored many of the first projects to combine thinking and convening systemically. Peter Woodrow invited me to partner with him in Burundi and has continued to work with Diana Chigas and Rob Ricigliano to apply systems thinking to peacebuilding. Batya Kallus invited me to design and lead the first Leveraging GrantMaking workshop. Andrew Jones, Don Seville, and Georgie Bishop offered me additional case material. Linda Booth Sweeney provided me with lots of encouragement and insight into what other colleagues are doing to "give system dynamics wings."

Thank you as well to Carol Gorelick for her valuable comments throughout, and to Michelle Heritage, R. Scott Spann, and Lisa Spinali for reviewing and enhancing parts of the manuscript.

None of this work happens without clients willing to engage in it. They and the people they serve are the ultimate reason this book was written. Thanks to those of you who have been so open to applying systems thinking to your work in making a better world, including: Joe Bartmann, Jennifer Bentley, Jacqueline Claunch, Beth Davis, Mark Draper, Nancy Frees Fountain, Jason Glass, Tim Jenkins, Karen Koenemann, Cynthia Lamberth,

Nancy Leonard, Jennifer Ludwig, Ann Mansfield, Anne Miskey, David Nee, John Sarisky, Bret Smith, David Tilly, Susan Tucker, Joan Wallace Benjamin, Jon Walz, and Traci Wodlinger.

I also want to acknowledge the many people who have embraced and contributed to this work through their participation in workshops on Leading Systemic Change and Leveraging Grant-Making in many different forums. I have learned from all of your insights and challenging questions.

Special thanks to Teri Behrens, editor-in-chief of *The Foundation Review*, for publishing an early version of these ideas and promoting their value in philanthropy. I also thank Joni Praded, my editor at Chelsea Green, for inviting me to write this book and providing enormously helpful feedback along the way. Thank you as well to Nancy Daugherty for her patience and persistence in developing the figures that are worth a thousand words.

Finally, none of this would have been written without the loving support of my wife, Marilyn Paul, an accomplished author in her own right and staunch believer in this work. She has been my closest sounding board over the years in helping me think through and explain what I do, and she offered many valuable ideas on the manuscript. I am also so grateful to my son, Jonathan Stroh, whose enthusiasm for life inspires me every day.

Oakland, CA
June 2015

Vicious Cycles of Climate Change

Increasing greenhouse gases or GHGs such as CO_2 in the earth's atmosphere creates several vicious cycles in nature that are likely to further increase global temperatures. These cycles are mapped out in figure A.1, offered here as an example of how to read a diagram showing several vicious cycles at work.

You'll see that as GHGs and temperatures increase, both soil and the oceans release more CO_2, as the Faster Release reinforcing loop (R) in

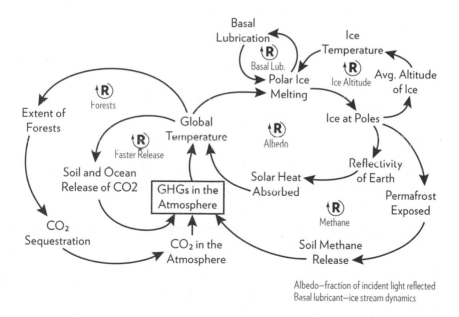

Albedo—fraction of incident light reflected
Basal lubricant—ice stream dynamics

FIGURE A.1 VICIOUS CYCLES OF CLIMATE CHANGE. Growing CO2 concentrations in the atmosphere have created vicious cycles in nature that are likely to increase global temperature over time. Adapted from Andrew Jones, Climate Interactive, 2015

figure A.1 depicts. Forest cover declines, which reduces the amount of CO_2 held or sequestered by trees, thereby increasing the level of CO_2 released into the atmosphere (R, Forests).

Moreover as temperatures increase, polar ice melts, which has other negative and amplifying impacts. When ice at the poles decreases, the earth becomes less reflective of the sun's rays, which increases the amount of solar heat absorbed, thereby increasing temperatures (R, Albedo). Ice melting also exposes permafrost, which releases methane, another greenhouse gas that is even more toxic than CO_2 (R, Methane). In addition, as ice at the poles decreases, the average altitude of the ice decreases, which reduces ice temperature and increases ice melt even faster (R, Ice Altitude). Finally, a potentially less influential impact of polar ice melting is basal lubrication, the flow of water produced by melting ice that increases the speed at which ice melts further (R, Basal Lubrication).

Sample Interview Questions for Specific Projects

When launching a systems-thinking project that involves multiple stake-holders, it is best to conduct interviews with project participants. You'll use the information you collect to build an initial systems map for discussion and input from the team. Below are the questions used in some of the cases discussed throughout this book.

B.1. QUESTIONS FOR THE AFTER PRISON INITIATIVE

1. Where within the overall field do you focus your work? Why do you focus here?
2. What other change efforts are you aware of in the field? How do these other efforts help/hinder your own? How do your efforts help—and potentially hinder—the work of others?
3. There are multiple barriers to reentry cited in the literature. In your mind, what are the three most important? *Why* do these barriers exist?
4. How do you engage other stakeholders, particularly those in related fields and with access to additional resources, in addressing these issues?
5. To what extent do you approach reentry as a larger issue around structural racism? For example, do you think in terms of enfranchisement as well as reentry for former prisoners, and/or the integration of entire marginalized communities into society? What end results are you aiming for? If you think in these larger terms, how explicit are you in doing so?

B.2. QUESTIONS FOR DEVELOPING
TEN-YEAR PLAN TO END HOMELESSNESS

1. What is the pattern of homelessness you've seen in Calhoun County over the past ten to twenty years?
2. What do you believe are the underlying causes of the problem?
3. What are the consequences of homelessness? How serious is the problem? What would happen over time if nothing else were done to address it?
4. What has your organization done to address the problem? What has been successful? What have you tried that has not been as successful as you hoped? Why do you think these shortfalls occurred?
5. What would you do differently? Why? What has prevented you from doing this?
6. No one wants homelessness to continue, yet it does. In your opinion, why does it persist?
7. In addition to what your organization can do, what else needs to be done to end homelessness in the county?
8. Who else at the state and local levels needs to be involved to contribute to the end of homelessness? What do they need to contribute—political leadership, funding, technical resources?

B.3. QUESTIONS FOR IMPROVING RURAL HOUSING

1. What is working now in your community to develop new housing?
2. What might you do to cultivate additional development?
3. What limits to appropriate, affordable housing development have you encountered?
4. How are you dealing with these limits now? How might you deal with them?
5. What limits might you encounter going forward?
6. Certain improvements in housing and other aspects of community development might take longer to realize than others. Where are the longest delays in your community? How can you

help people shorten those delays or increase their patience in staying the course?

7. There tends to be an understandable focus on single-family unattached homes. What place, if any, do you see for multifamily housing or contiguous single-family housing (such as town houses)?

B.4. QUESTIONS FOR COLLABORATING FOR IOWA'S KIDS

Questions to Clarify Prospects for Partnership

1. How do you want to define IDE success? What should be its unique contributions?
2. How do you want to define AEA System success? What should be its unique contributions?
3. How will IDE's success contribute to AEA System success?
4. How will AEA System success contribute to IDE success?

Questions to Clarify Accidental Adversaries Relationship

1. When IDE has problems achieving results, what does it do to get back on track?
2. How do these actions unwittingly undermine AEA System success?
3. When the AEA System has problems achieving results, what does it do to get back on track?
4. How do these actions unwittingly undermine IDE success?

Multiple-Archetype Diagrams

These diagrams show how multiple archetypes can illuminate a single complex problem. In the first diagram, C.1, the Shifting the Burden dynamic adds to and complements the Fixes That Backfire dynamic of mass incarceration in figures 7.1 and 7.2. Figures C.2 through C.4 show different aspects of the deeply rooted problem of identity-based conflicts, where both parties to a conflict believe that their very rights to exist are threatened. C.2 describes the Shifting the Burden archetype underlying these conflicts. C.3 illuminates the challenge of Conflicting Goals. Finally, C.4 shows how these conflicts embed themselves even further through the dynamic of Escalation. For readers interested in learning more about the system dynamics of identity-based conflicts and recommended interventions to resolve them, go to the reference noted on these diagrams.

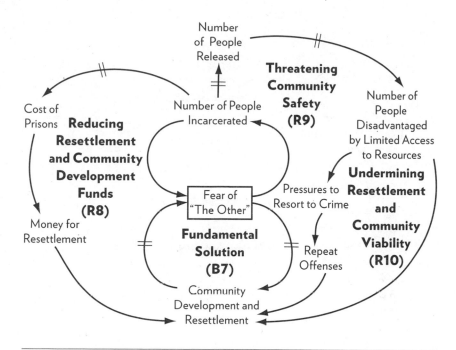

FIGURE C.1 AN ADDICTION TO PRISON. Depending on mass incarceration to reduce people's fear of "the other" undermines society's ability to implement the more fundamental solutions of community development and resettlement. The high costs of prison reduce the money available to spend on these responses (R8). Disadvantages experienced by formerly incarcerated people often lead to repeat offenses, which destabilize the community even further (R9). Even if these people do not commit another crime, it is difficult for them to become net contributors to the community's resources (R10). Modified from a diagram developed by Seed Systems for Open Society Institute

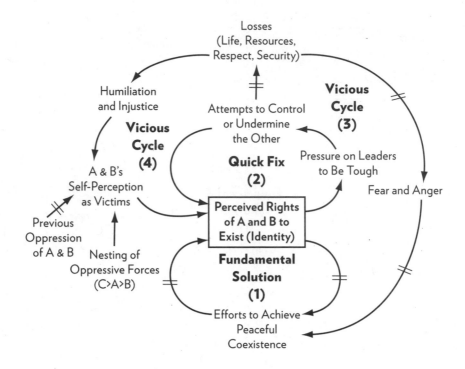

FIGURE C.2 IDENTITY-BASED CONFLICTS: SHIFTING THE BURDEN. In order to demonstrate their respective rights to exist, people on both sides of an identity-based conflict tend to become addicted to trying to control or undermine people on the other side of this conflict. The stronger party (A)—economically and militarily—seeks to exert control while the weaker side (B) works to undermine the stronger one. David Peter Stroh, "The System Dynamics of Identity-Based Conflict," in D. Korppen, N. Ropers, and H. J. Giessmann (editors), *The Non-Linearity of Peace Processes* (Barbara Budrich Publishers, 2011)

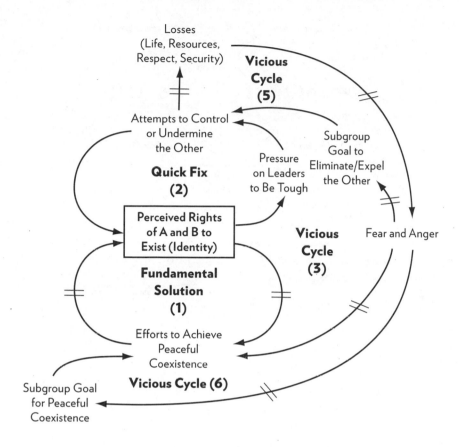

FIGURE C.3 IDENTITY-BASED CONFLICTS: CONFLICTING GOALS. Each side also has subgroups that seek to resolve the conflict through peaceful coexistence rather than complete victory. However, the subgroups in favor of complete victory tend to dominate over time, since their more aggressive tactics generate emotions of fear and anger that are more viscerally compelling than those of peace. Stroh, "The System Dynamics of Identity-Based Conflict"

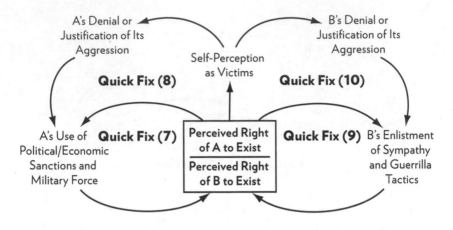

FIGURE C.4 IDENTITY-BASED CONFLICTS: ESCALATION. These problems are compounded because both sides become caught in an endless race for both domination and victimization since oppressing from the victim position is often viewed as a more legitimate stance. Stroh, "The System Dynamics of Identity-Based Conflict"

Additional Resources

The following compilation of books and websites summarizes: diverse approaches to thinking and acting systemically; applications of systems thinking to such challenges as urban renewal, peacebuilding, sustainable development, K–12 education, healthcare, and ending homelessness; and networks of people and archives of materials dedicated to supporting the use of systems thinking nationally and around the world.

BOOKS

Linda Booth Sweeney:
> *When a Butterfly Sneezes*, Pegasus Communications, 2001. A guide for helping kids (of all ages) explore interconnections in our world through favorite stories.
> *The Systems Thinking Playbook*, co-authored with Dennis Meadows, Chelsea Green, 2010. A handbook of experiential exercises that stretch and build learning and systems thinking capabilities.

Peter Checkland and Jim Scholes, *Soft Systems Methodology in Action*, Wiley, 1999. Offers an alternative systems thinking methodology that is not based on the visual language of causal loop diagramming.

Glenda Eoyang and Royce Halladay, *Adaptive Action: Leveraging Uncertainty in Your Organization*, Stanford University Press, 2013. Offers another systems thinking methodology that is not based on the visual language of causal loop diagramming.

Jay Forrester, *Urban Dynamics*, Pegasus Communications, 1969. A classic in the field of system dynamics that explains the shortcomings of urban renewal strategies employed in the 1960's United States War on Poverty.

D. Korppen, N. Ropers, H.J. Giessmann (eds.), *The Non-Linearity of Peace Processes*, Barbara Budrich Publishers: Farmington Hills, MI, 2011. A collection of papers that apply different methods of thinking systemically to the challenge of peace-building.

Donella Meadows:

Limits to Growth, Universe Books, 1972. A system dynamics model developed for the Club of Rome describes the likely environmental impacts of unbridled population and economic growth.

Limits to Growth: The 30-Year Update, co-authored with Jørgen Randers and Dennis Meadows, Chelsea Green, 2004. This 30-year follow-up report builds on many of the predictions made in the original report and recommends how to avoid environmental collapse.

Thinking in Systems, Chelsea Green, 2008. An outstanding introduction to systems thinking by a pioneering leader in the field.

C. Otto Scharmer, *Theory U: Leading From the Future as It Emerges*, Berrett-Koehler, 2009. Describes a comprehensive framework for managing complex deep change that is complementary to the one introduced in this book.

Peter Senge, et al.:

The Fifth Discipline, Doubleday, 1990, second edition published in 2006. The pioneering framework underlying many of the ideas in this book. Systems thinking is "the fifth discipline."

The Fifth Discipline Fieldbook, co-authored with Art Kleiner, Charlotte Roberts, Richard Ross, and Bryan Smith, Doubleday, 1994. Grounds the principles of *The Fifth Discipline* in highly accessible strategies and tools that enable readers to apply each of the organizational learning disciplines.

The Necessary Revolution, co-authored with Bryan Smith, Nina Kruschwitz, Joe Laur, and Sara Schley, Broadway Books, 2008. Applies the five disciplines of organizational learning to building a sustainable world.

Schools That Learn (Updated and Revised), co-authored with Nelda Cambron-McCabe, Timothy Lucas, Bryan Smith, Janis Dutton, and Art Kleiner, Crown Business, 2012. Applies the five disciplines of organizational learning to meeting the challenge of K–12 education.

WEBSITES

Applied Systems Thinking, www.appliedsystemsthinking.com. An informational website on applied systems thinking developed by Michael Goodman and David Peter Stroh.

Bridgeway Partners, www.bridgewaypartners.com. Materials and services offered by David Peter Stroh and Marilyn Paul to help people reframe and resolve intractable problems—including a range of social issues and the challenges of individual and organizational overload.

Climate Interactive, www.climateinteractive.org. Provides computer simulations and other learning materials covering such topics as climate change, clean energy, disaster risk reduction, and resilience.

Creative Learning Exchange, www.clexchange.org. Nonprofit founded by Jay Forrester to introduce system dynamics into the K–12 school curriculum.

Funders Together to End Homelessness, www.funderstogether.org. A national network of funders committed to ending and preventing homelessness using a systems approach.

International System Dynamics Society, www.systemdynamics.org. An international, nonprofit organization devoted to encouraging the development and use of System Dynamics and systems thinking.

ReThink Health, www.rethinkhealth.org. Develops and compiles practical information, tools, and approaches to support leaders committed to improving regional healthcare delivery—including through its nationally recognized ReThink Health Dynamics Model.

The Systems Thinker, www.thesystemsthinker.com. New website includes the complete library of 800+ Systems Thinker articles as well as a number of guides and webinars. Each article is free and available in an easy-to-read, interactive format as well as the original PDFs.

Systems Thinking World, www.systemswiki.org. An online community dedicated to creating content, fostering interactions, and applying systems thinking to further an understanding of a systemic perspective and enable thinking and acting systemically. Also provides a comprehensive online learning environment at https://kumu.io/stw/systems-kele.

NOTES

INTRODUCTION

1. See, for example, Peter Senge, *The Fifth Discipline* (updated and revised; Doubleday, 2010) and Peter Senge et al., *The Fifth Discipline Fieldbook* (Doubleday, 1994).
2. See, for example, Peter Checkland and Jim Scholes, *Soft Systems Methodology in Action* (Wiley, 1999).
3. *Harvard Business Review* staff, "You Can't Be a Wimp: Make the Tough Calls" (interview with Ram Charan), *Harvard Business Review*, November 2013.
4. See, for example, Otto Scharmer, *Theory U: Leading from the Future as It Emerges* (Berrett-Koehler, 2009); John Kania and Mark Kramer, "Collective Impact," *Stanford Social Innovation Review*, Winter 2011; and Zaid Hassan, *The Social Labs Revolution* (Berrett-Koehler, 2014).
5. This insight has been attributed to many systems thinkers, including Dr. Paul Batalden, professor emeritus at the Geisel School of Medicine at Dartmouth College; Don Berwick, past president and CEO of the Institute for Healthcare Improvement and former administrator of the US Centers for Medicare and Medicaid Services; and Edward Deming, founder of the quality movement.

CHAPTER 1

1. Matthew B. Durose, Alexia D. Cooper, and Howard N. Snyder, "Recidivism of Prisoners Released in 30 States in 2005: Patterns from 2005 to 2010," US Department of Justice, Office of Justice Programs, Bureau of Justice Statistics, April 2014.
2. Lewis Thomas, *The Medusa and the Snail* (Viking, 1979).
3. Donella Meadows, *Thinking in Systems* (Chelsea Green, 2008).
4. For a broader understanding of the different schools of systems thinking and when to apply which methodology, read Michael C. Jackson, *Systems Thinking: Creative Holism for Managers*, Wiley & Sons, 2003.
5. Senge, *The Fifth Discipline*, and Senge et al., *The Fifth Discipline Fieldbook*. Causal loop diagrams were originally documented by my colleague

Michael Goodman in his pioneering textbook *Study Notes in System Dynamics* (1974, reprinted by Pegasus Communications in 1989).

CHAPTER 2

1. Carol Dweck, *Mindset: The New Psychology of Success* (Random House, 2006).
2. Senge, *The Fifth Discipline*, p. 231.
3. For more information, you can read David Peter Stroh, "The System Dynamics of Identity-Based Conflicts," in D. Korppen, N. Ropers, and H. J. Giessmann (editors), *The Non-Linearity of Peace Processes* (Barbara Budrich Publishers, 2011).
4. John Kania and Mark Kramer, "Collective Impact," *Stanford Social Innovation Review*, Winter 2011.
5. Paul Schmitz, "The Real Challenge for Collective Impact," *Huffington Post*, September 27, 2012.

CHAPTER 3

1. OSI has since changed its name to Open Society Foundations.
2. Jonathan Simon, "Governing Through Crime: How the War on Crime Transformed American Democracy and Created a Culture of Fear," *Studies in Crime and Public Policy*, 2009.
3. Buckminster Fuller and others have observed that the underlying problem faced by the world is not limited resources but the ineffective distribution of sufficient resources.
4. Marvin Weisbord, *Productive Workplaces: Dignity, Meaning, and Community in the 21st Century* (Jossey-Bass, 1991).
5. Alfred Blumstein and Joel Wallman, *The Crime Crop in America* (Cambridge University Press, 2005).
6. Steven D. Levitt, "Understanding Why Crime Fell in the 1990s: Four Factors That Explain the Decline and Six That Do Not," *Journal of Economic Perspectives* 18, no. 1 (Winter 2004).
7. Wikipedia, http://en.wikipedia.org/wiki/Abracadabra.
8. Because not all nouns are variables, a good way to confirm if a factor varies is to put the words *level of* in front of it: Level of Repeat Offenses. For example, the Number of People Incarcerated is a variable because it can increase or decrease over time. However, Strategy is not a variable, although Relative Emphasis on Punishment Versus Reform is a variable because that emphasis can change over time.
9. Peter Woodrow, "Conflict Analysis of Burundi" (CDA Collaborative Learning Projects, 2004).
10. It is also possible to substitute colored links—say, green for "same" and red for "opposite"—on final diagrams.

CHAPTER 4

1. Meadows, *Thinking in Systems*.
2. Jim Collins, *Good to Great* (HarperCollins, 2001) and *Good to Great and the Social Sectors* (monograph; HarperCollins, 2005).
3. David Peter Stroh, "A Systems View of the Economic Crisis," *The Systems Thinker* 20, (February 2009).
4. P. Ball, *Critical Mass: How One Thing Leads to Another* (Farrar, Straus and Giroux, 2006); Malcolm Gladwell, *The Tipping Point* (Little Brown, 2000).
5. D. H. Meadows et al., *The Limits to Growth* (Universe Books, 1972).
6. G. Kelling and J. Q. Wilson, "Broken Windows," *The Atlantic* 249, no. 3 (March 1982).
7. C. Johnson, "Study Backs Theory That Links Conditions, Crime," *Boston Globe*, February 8, 2009.
8. J. A. Fox, "Ganging Up," *Boston Globe*, December 1, 2003.
9. E. Moscowitz, "Just Say 'In the Know,'" *Boston Globe*, December 22, 2008.
10. Institute on Education and the Economy, "Using What We Have to Get the Schools We Need: A Productivity Focus for American Education," IEE Document No. Bk-5 (Teachers College, Columbia University, October 1995). I wish to thank my colleagues Jennifer Kemeny and Sherry Immediato, who worked on this case and brought it to my attention.
11. The pioneering work on codifying system archetypes and making them accessible was done by Michael Goodman, Jennifer Kemeny, and Charlie Kiefer at Innovation Associates. Most of these are summarized in William Braun, *The System Archetypes*, 2002, downloadable from http://www.albany.edu/faculty/gpr/PAD724/724WebArticles/sys_archetypes.pdf.
12. S. Friedman, "When Heroin Supply Cut, Crime Rises, Says Report," *Boston Globe*, April 22, 1976.
13. Linda Polman, *The Crisis Caravan* translated by Liz Waters (Metropolitan Books, 2010).
14. William Easterly, *The Tyranny of Experts* (Basic Books, 2014).
15. In their recent book *A Path Appears* (Knopf, 2014) award-winning journalist Nicolas Kristof and his wife, Sheryl WuDunn, offer many excellent examples of bottom-up development.
16. John R. Ehrenfeld, "The Roots of Sustainability," *Sloan Management Review*, 46, no. 2 (Winter 2005).
17. Peter Buffett, "The Charitable-Industrial Complex," *New York Times*, July 27, 2013.
18. *The Dance of Change*, a book by Peter Senge et al. (Doubleday Currency, 1999), uses the Limits to Growth archetype to explain the obstacles to organizational change and what can be done to overcome them.
19. Meadows et al., *Limits to Growth*.

20. Thomas Piketty, *Capital in the 21st Century*, translated by Arthur Gold-hammer (President and Fellows of Harvard College, 2014).

21. Peter Stroh, "The Rich Get Richer, and the Poor . . .", *The Systems Thinker*, March 1992.

22. Keith Lawrence and Terry Keleher, *Structural Racism* (Race and Public Policy Conference, 2004).

23. Nicholas Kristof and Sheryl WuDunn, "The Way to Beat Poverty," *New York Times*, September 14, 2014.

24. Adapted from copyrighted work on "The Leadership Dilemma in a Democratic Society," developed by and reproduced with permission of the Public Sector Consortium, 2003. Learn more at http://www.public-sector .org/file/The-Leadership-Dilemma-in-Democratic-Society.pdf.

25. Peter Stroh, "Conflicting Goals: Structural Tension at Its Worst," *The Systems Thinker*, September 2000.

26. David Peter Stroh, "A Systemic View of the Israeli–Palestinian Conflict," *The Systems Thinker*, June–July 2002.

27. David Peter Stroh, "The System Dynamics of Identity-Based Conflict," in Korppen, Ropers, and Giessmann (editors), *The Non-Linearity of Peace Processes.*

28. Terrence Real, *The New Rules of Marriage* (Ballantine Books, 2007).

29. Elinor Ostrom, *Governing the Commons: The Evolution of Institutions for Collective Action* (Cambridge University Press, 1990).

30. *National Geographic*, "The Carbon Bathtub," December 2009.

31. Levitt, "Understanding Why Crime Fell in the 1990s."

CHAPTER 5

1. David Peter Stroh and Michael Goodman, "A Systemic Approach to Ending Homelessness," Applied Systems Thinking Journal, October 2007. Download from: http://www.appliedsystemsthinking.com/supporting _documents/TopicalHomelessness.pdf.

2. Numbers derived from Calhoun County 2013–2014 Report Card, The Coordinating Council, http://www.tcccalhoun.org/CRC/2013-14%20 TCC%20Report%20Card.pdf.

3. Marvin Weisbord and Sandra Janoff, *Future Search*, 3rd edition, Berrett-Koehler, 2011; Harrison Owen, *Open Space Technology*, 3rd edition, Berrett-Koehler, 2008; Juanita Brown, *The World Café*, Berrett-Koehler, 2005.

4. Ram Nidumolu et al., "The Collaboration Imperative," *Harvard Business Review*, April 2014; Scharmer, *Theory U*; Zaid Hassan, *The Social Labs Revolution*, Reos Publications, 2014.

5. Senge, *The Fifth Discipline*, pp. 150–55.

6. Scharmer, *Theory U*, p. 134.

CHAPTER 6

1. Kathleen Zurcher and Timothy Grieves, "Collaborating for Iowa's Kids," Iowa Department of Education and Iowa Area Education Agencies, August 17, 2012.
2. Shirley Leung, "Pine St. Inn's Bold Move to End Chronic Homelessness," *Boston Globe*, July 16, 2014.
3. Balancing advocacy and inquiry—along with many other tools to facilitate team learning—is described in Peter Senge et al., *The Fifth Discipline Fieldbook*.
4. Adapted from a sequence developed by Cliff Barry, Shadow Work Seminars.
5. Scharmer, *Theory U*.
6. Real, *The New Rules of Marriage*.

CHAPTER 7

1. Durose, Cooper, and Snyder, "Recidivism of Prisoners Released in 30 States in 2005."
2. Charles M. Blow, "Crime and Punishment," *New York Times*, December 1, 2014.
3. Ilya Somin, "Conservatives Rethinking Mass Imprisonment and the War on Drugs," *Washington Post*, January 26, 2014. One interesting example of this shift is the recent formation of a highly unlikely coalition to fix the criminal justice system that includes organizations from both the political right and left; to learn more about this new Coalition for Public Safety, go to www.coalitionforpublicsafety.org.
4. Kristof and WuDunn, *A Path Appears*.
5. Ibid.
6. For more on system dynamics modeling related to public policy, you can read "How Small System Dynamics Models Can Help the Public Policy Process" by Navid Ghaffarzadegan, John Lyneis, and George P. Richardson, a white paper published by the Rockefeller College of Public Affairs and Policy, University of Albany, SUNY, available at http://www.albany.edu /~gpr/SmallModels.pdf.

CHAPTER 8

1. Adding mental models to causal loop diagrams was a technique developed by Innovation Associates Organizational Learning.

CHAPTER 9

1. Robert Kegan and Lisa Laskow Lahey, *How We Talk Can Change the Way We Work* (Jossey-Bass, 2001).
2. This insight has been attributed to many systems thinkers, including Dr. Paul Batalden, professor emeritus at the Geisel School of Medicine at Dartmouth College; Don Berwick, past president and CEO of the Institute for Healthcare Improvement and the former administrator of the US Centers for Medicare and Medicaid Services; and Edward Deming, founder of the quality movement.
3. Peter Stroh and Wynne Miller, "Learning to Thrive on Paradox," *Training and Development*, September 2014.
4. Leung, *Pine St. Inn's Bold Move to End Chronic Homelessness*.
5. Scharmer, *Theory U*.
6. Everett M. Rogers, *Diffusion of Innovations*, 5th edition (Free Press, 2003).

CHAPTER 10

1. David Peter Stroh and Marilyn Paul, "Is Moving Too Fast Slowing You Down? How to Prevent Overload from Undermining Your Organization's Performance," *Reflections: The Society for Organizational Learning Journal* 13, no. 1 (Summer 2013).
2. Meadows, *Thinking in Systems*.
3. Barbara Tuchman, *The March of Folly* (Random House, 1984).
4. Claudia Dreifus, "A Chronicler of Warnings Denied," *New York Times*, October 28, 2014.
5. Leung, *Pine St. Inn's Bold Move to End Chronic Homelessness*.
6. Pathways to Housing website at https://pathwaystohousing.org/housing-first-model.
7. David Peter Stroh, *A Systems Approach to Improving Environmental Public Health*, 2013. Unpublished report prepared for the Centers for Disease Control; copies available from the author.
8. David Cooperrider and Diana Whitney, *Appreciative Inquiry: A Positive Revolution in Change* (Berrett-Koehler, 2005); Richard Pascale, Jerry Sternin, and Monique Sternin, *The Power of Positive Deviance* (Harvard Business Press, 2010).
9. David Peter Stroh and Marilyn Paul, "Managing Your Time as a Leader," *Reflections: The Society for Organizational Learning Journal*, Winter 2006.
10. Leung, *Pine St. Inn's Bold Move to End Chronic Homelessness*.
11. Case study developed by David Peter Stroh and John McGah for Funders Together to End Homelessness, a national association of funders committed to ending homelessness. Contact Funders Together at www.funderstogether.org for copies. Michelle Heritage, executive director

of CSB, was especially helpful in providing information about the organization.

12. Kania and Kramer, "Collective Impact."

13. The tools as defined here were developed by Innovation Associates, though other versions of both exist.

14. Meadows, *Thinking in Systems*, p. 164.

15. Reflecting on Peace Practice Program, *Key Principles in Effective Peace-building* (CDA Collaborative Learning Projects, 2014).

16. To learn more, contact Pamela Wilhelms at the Wilhelms Consulting Group, http://wcgsite.weebly.com/about.html.

17. Grantmakers for Effective Organizations, *2013 Pathways to Grow Impact: Philanthropy's Role in the Journey* (GEO Resource Library, January 29, 2013).

18. Reflecting on Peace Practice Program, *Lessons from Program Effectiveness*.

19. Kristof and WuDunn, *A Path Appears*.

CHAPTER 11

1. Michael Goodman and Art Kleiner, "The Archetype Family Tree," in Peter Senge et al., *The Fifth Discipline Fieldbook*.

2. Another example of a planning approach for social innovation is design thinking as summarized by Tim Brown and Jocelyn Watt, "Design Thinking for Social Innovation," *Stanford Social Innovation Review*, Winter 2010.

3. See, for example, Cooperrider and Whitney, *Appreciative Inquiry*; and Pascale, Sternin, and Sternin, *The Power of Positive Deviance*.

4. The reader is encouraged to review chapter 9 in situations where it is difficult to develop alignment around a vision for the system.

5. Guidelines for scaling up are summarized in chapter 10 and can also be found at Grantmakers for Effective Organizations, *2013 Pathways to Grow Impact: Philanthropy's Role in the Journey* (GEO Resource Library, January 29, 2013).

6. Financial estimate based on research by the Massachusetts Housing and Shelter Alliance.

7. David Peter Stroh and Kathleen Zurcher, "Leveraging Grantmaking—Part 2: Aligning Programmatic Approaches with Complex System Dynamics," The Foundation Review, Winter 2010.

8. Daniel H. Kim, *Organizing for Learning* (Pegasus Communications, 2001).

9. David Peter Stroh, "What to Do When You Have Too Many Goals," blog post, http://www.bridgewaypartners.com/Blog/tabid/67/entryid/17/What-to-Do-When-You-Have-Too-Many-Goals.aspx.

10. To learn more about computer simulations that address climate change and sustainability, look at the C-ROADS simulation at www.climateinteractive.org

and the Threshold 21 (T21) model at http://millennium-institute.org. To learn more about computer simulations for health care delivery, read the work of Gary Hirsch, beginning with the following article: Bobby Milstein, Jack Homer, and Gary Hirsch, "Analyzing National Health Reform Strategies with a Dynamic Simulation Model," *American Journal of Public Health*, May 2010.

CHAPTER 12

1. Stroh and Paul, "Is Moving Too Fast Slowing You Down?"
2. Ibid.
3. American Productivity and Quality Center, *White Collar Productivity Improvement* (APQC, 1986).
4. *The Water of Ayole*, video directed by Sandra Nichols, available for viewing at http://vimeo.com/6281949. This twenty-seven-minute video can also serve as an excellent introduction to the differences between conventional and systems thinking. The first half (through 13:43) exemplifies the former and the second half, the latter.
5. Kristof and WuDunn, *A Path Appears*, pp. 167–231.

CHAPTER 13

1. Senge, *The Fifth Discipline*, pp. 374–75.
2. Meadows, *Thinking in Systems*, p. 167.
3. Marilyn Paul, "Moving from Blame to Accountability," *The Systems Thinker* 8, no. 1 (1997).
4. David Peter Stroh, "The Systems Orientation: From Curiosity to Courage," *The Systems Thinker* 21, no. 8 (2011).
5. To learn more, go to http://www.corechangecincy.com.
6. You can learn more about the online program and preview it at http://www.iseesystems.com/store/Training/ApplySysThink.aspx.

INDEX

ABOUT THE AUTHOR

Kim Kennedy

David Peter Stroh is a founding partner of Bridgeway Partners and a founding director of the website Applied Systems Thinking. He also cofounded Innovation Associates, the consulting firm whose pioneering work in the area of organizational learning formed the basis for fellow cofounder Peter Senge's management classic *The Fifth Discipline.*

Internationally recognized for enabling people to apply systems thinking to achieve breakthroughs around chronic, complex problems, David has worked with organizations and communities across the nonprofit, private, and public sectors to develop social-change initiatives that improve system-wide performance over time. His clients have included the W.K. Kellogg Foundation, the Open Society Foundations, the Alliance for International Conflict Prevention and Resolution, the National Center on Family Homelessness, the national Centers for Disease Control, the World Bank, and Royal Dutch Shell, among others.

David also co-created and leads the Leading Systemic Change workshop, is a charter member of the Society for Organizational Learning, and is a frequent speaker on systems thinking, organizational learning, and social change. His writing has appeared in numerous professional books and journals, including *Transforming Work, Reflections: The Society for Organizational Learning Journal, The Systems Thinker,* and *OD Practitioner.* David was a National Science Foundation fellow at MIT where he received a master's in city planning, and he graduated summa cum laude with undergraduate degrees in civil engineering and urban studies from the University of Michigan.

For more information, please visit www.bridgewaypartners.com and www.appliedsystemsthinking.com.

green press
INITIATIVE

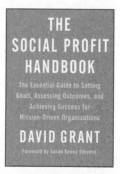

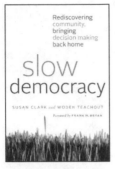
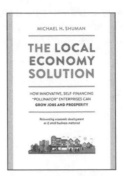